BOVINE ENDOSCOPY

Sotirios Karvountzis

DVM MRCVS

CABI is a trading name of CAB International

CABI	CABI
Nosworthy Way	200 Portland Street
Wallingford	Boston
Oxfordshire OX10 8DE	MA 02114
UK	USA
Tel: +44 (0)1491 832111	Tel: +1 (617)682-9015
E-mail: info@cabi.org	E-mail: cabi-nao@cabi.org
Website: www.cabi.org	

The views expressed in this publication are those of the author(s) and do not necessarily represent those of, and should not be attributed to, CAB International (CABI). Any images, figures and tables not otherwise attributed are the author(s)' own. References to internet websites (URLs) were accurate at the time of writing.
CAB International and, where different, the copyright owner shall not be liable for technical or other errors or omissions contained herein. The information is supplied without obligation and on the understanding that any person who acts upon it, or otherwise changes their position in reliance thereon, does so entirely at their own risk. Information supplied is neither intended nor implied to be a substitute for professional advice. The reader/user accepts all risks and responsibility for losses, damages, costs and other consequences resulting directly or indirectly from using this information.

CABI's Terms and Conditions, including its full disclaimer, may be found at https://www.cabi.org/terms-and-conditions/.

A catalogue record for this book is available from the British Library, London, UK.

ISBN-13: 9781789246667 (paperback)
        9781789246674 (ePDF)
        9781789246681 (ePub)

DOI: 10.1079/9781789246681.0000

Commissioning Editor: Alexandra Lainsbury
Editorial Assistant: Lauren Davies
Production Editor: Rosie Hayden

Typeset by Straive, Pondicherry, India
Printed and bound in the UK by Severn, Gloucester

BOVINE ENDOSCOPY

To Helen

    To Athena

        To Sofia-Maria

Thank you for your selfless sacrifices and the principles you instilled in me

Contents

Author Biography

Sotirios Karvountzis

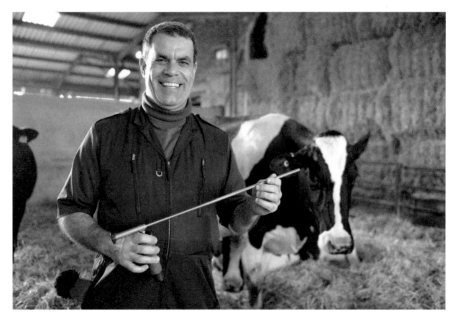

I qualified in 1994, from the Veterinary School of the Aristotle University of Thessalonica, Greece.

My main clinical interests are advanced reproductive technologies (ART) for cattle and sheep, with multiple ovulation and embryo transfer (MOET) in cattle and sheep, cattle ovum pick-up (OPU) and *in-vitro* fertilization (IVF), as well as genomic benchmarking, reproductive and abdominal surgery, hoof health, mastitis, bull and ram semen testing, youngstock health, pathology and ruminant nutrition, based on Cornell Net Carbohydrate Protein System (CNCPS).

I have extensive experience in training vets and farmers, as a CowSignals Master Trainer and a LANTRA Freelance Instructor and Assessor. I have trained most of the vets who practice endoscopic surgery in the UK, and I have tutored endoscopic courses in the UK, Europe, Australia and North America.

I am the tutor for a cattle DIY artificial insemination course and a non-vet cattle scanning course. I organize foot trimming courses and hold the Dutch Diploma in Cattle Hoofcare, the NPTC Level 3 Certificate of Competence in Cattle Foot Trimming, White Line Atlas Hoof Trimming Instructor, to mention a few.

I have extensive experience on improving udder health, with particular interest in dynamic and static milk parlour testing, aiming to identify risk factors for mastitis during milking.

I am deeply interested in clinical research and use Stata, by StataCorp LLC, to support my statistical analysis. I coordinated two large-scale clinical research projects, on caesarean sections and on left displacement of abomasum corrective techniques.

I am also a computer programmer, with particular interest in SQL databases. I developed the analysis program MilkMonitor©, which can analyse large amounts of cattle data.

I also hold a helicopter commercial pilot's Licence CPL(H) with instrument rating IR(H).

1 Introduction

Bovine endoscopy is a surgical technique with a unique place in veterinary medicine, permitting minimally invasive examination and treatment of various ailments. The word endoscopy is derived from the Greek words *endon* (Ἐνδον) for 'internal' and *skopein* (Σκοπεῖν), for 'inspect'.

Introducing endoscopic techniques revolutionizes applications of bovine medicine by facilitating farm animal veterinary surgeons to examine, establish a diagnosis and treat various conditions without the need for open and intrusive surgery.

From explorations of the abdomen and the thorax to abdominal corrections for left and right displacement of abomasum, bovine endoscopy has a wide range of applications and offers a multitude of advantages over traditional diagnostic and surgical techniques.

Endoscopy finds application in several areas of veterinary medicine. Some common indications for endoscopic procedures in cattle include the evaluation of left and right displacement of the abomasum, traumatic reticulitis, intestinal ulcers, peritonitis, kidney conditions, chronic gastrointestinal disorders, and the investigation of respiratory ailments, including pneumonia and pleuritis.

Additionally, this technique is crucial as it allows the examination of the uterus as it lies in the retroperitoneal space, theloscopy and corrective procedures of the teat, and the collection of samples for diagnostic purposes.

Why Endoscopy Is Important

Endoscopy provides a direct and clear examination of the areas of interest, allowing for precise diagnosis and effective treatment.

By accessing the abdomen, the thorax or the teat, endoscopy enables veterinarians to identify abnormalities, such as displacements, chronic inflammation, foreign bodies, ulcers and other abnormalities that may otherwise go undetected.

© Sotirios Karvountzis 2023. *Bovine Endoscopy* (Sotirios Karvountzis)
DOI: 10.1079/9781789246681.0001

Furthermore, bovine endoscopy offers a less aggressive substitute to open surgery, for example reducing risks that stem from laparoscopy, such as post-operative complications and prolonged recovery time. Due to its less invasive nature, endoscopy presents new diagnostic alternatives to areas of the body that were previously considered out of bounds, such as the thorax.

The ability to perform procedures such as biopsies, foreign body removals and surgical corrections not only enhances the welfare of the animal, but significantly improves the repertoire of skills of the veterinary surgeon.

Instrumentation and Techniques

Bovine endoscopy utilizes surgical instruments, including rigid endoscopes and increasingly flexible ones, that are designed to access the abdomen, thorax and teat of cattle. Endoscopes generally used in farm animal practice are almost exclusively rigid endoscopes.

While flexible endoscopes are also used as a matter of routine in other disciplines of veterinary medicine, such as equestrian and companion animal medicine, a reason that they are not commonplace in ruminant surgery is cost. Such investment is currently cost prohibitive in most cases, particularly if the operation is expected to be performed away from the veterinary practice premises, where the working conditions are unpredictable or the support staffing may not be available.

There are substantial considerations in using flexible endoscopes in veterinary medicine. A flexible endoscope can reach beyond the longest rigid endoscope into organs with considerable length. The commonest use of flexible endoscopes is for gastrointestinal examinations, from the colon to the oesophagus and the gastro-duodenum. Other organs that are frequently examined with the flexible endoscope are the trachea and the bronchi, naso-pharynx and the male urethra all the way up to the urinary bladder.

As the cost of the endoscopic instruments is constantly reducing, the author expects further future innovations, the primary one being the combining of portals, mainly that of the optical and working ones into one portal. Combined endoscopic instruments that include a working channel, which are already used in human medicine for procedures such as the collection of biopsies and the removal of foreign bodies, are slowly being integrated into large-animal veterinary medicine too.

Most flexible endoscopes used in veterinary medicine are combined instruments, and some rigid endoscopes are combined too. Although a combined rigid endoscope is likely to be more expensive than a single-channel rigid instrument of the same length and width, its use in veterinary medicine brings numerous advantages. For example, in shallow-depth exploratory operations half the number of portals are required, so the intrusion to the patient is minimized further.

Endoscopy is now an accepted and commonly used diagnostic and corrective procedure in human medicine, and there is much that we as veterinary surgeons can learn from the benefits of this minimally invasive technique, not least of which is the considerable acceleration it allows in patient recovery time. Bovine endoscopy offers a less invasive alternative to open surgery, reducing the associated risks, post-operative complications, and recovery time for cattle.

If a veterinary procedure, for instance the correction of a left displaced abomasum (LDA), can be carried out simply by making a number of 1–2 cm incisions known as portals and introducing a slender rigid or flexible instrument to visualize the displaced abomasum and correct the condition, we can surely no longer justify cutting open the body for diagnosis or correction. The technique also allows us to carry out abdominal explorations to identify concurrent disease such as peritonitis, abscess or the presence of foreign bodies.

In conventional surgery, even the most gentle of veterinary surgeons can cause immense trauma, even by the mere act of moving the intestine to the right or the left. In human medicine, this level of insult to the body can leave the patient unable to eat by mouth for six or more weeks. Slow recovery time is seen in ruminants too, impacting on production and welfare. Yet here we have a procedure at our disposal with a recovery time measured in days, not weeks.

Let us consider the uses for endoscopy in bovine medicine and its advantages and disadvantages.

Endoscopy has the following applications:

- LDA corrections;
- right displaced abomasum (RDA) corrections;
- exploratory laparoscopy;
- exploratory thoracoscopy; and
- teat endoscopy (theloscopy).

Advantages of endoscopy include the following:

- Allows full visualization of the target organs.
- LDA cows that are endoscopically corrected will produce more milk post-operatively than their laparotomically corrected counterparts.
- Speedier clinical recovery and a quicker return to normal yields.
- Minimal intrusion and infection risk.
- Animal can join the herd at the next milking.
- One-step (standing) techniques require one assistant only.
- Unique selling point for veterinary practices that apply the technique.

Disadvantages of endoscopy include the following:

- Purchase cost of the kit is equivalent to the cost of a good-quality ultrasound scanner.
- For a multi-branch practice with a limited number of surgeons trained in the technique, day-to-day management of the endoscopic equipment can present a challenge. The best results are achieved when there is at least one endoscopic kit allocated per branch and a minimum of two trained surgeons in each branch.

When these considerations are weighed up, there is a strong case for the veterinary profession to embrace endoscopy and make its use commonplace, to diagnose and treat conditions without subjecting our patients to unnecessary stress.

This book is a guide to bovine endoscopy, from explorations of the abdomen and the thorax to abdominal corrections for LDAs and RDAs. It also considers the rapidly expanding application of teat endoscopy – a procedure with minimal intrusion and few post-operative complications.

Endoscopic corrective techniques can be classified in two broad categories:

- the one-step technique, where preparation and correction take place while the animal is standing; and
- the two-step technique, where preparation takes place while the animal is standing and correction while in dorsal recumbency.

The one-step technique includes two applications, first where the abomasopexy point lies caudally to the xiphoid process, used for LDA corrections, and second, where the abomasopexy point lies cranially to the navel, also used for LDA corrections. The two-step technique, where the abomasopexy point lies cranially to the navel, is used for LDA and RDA corrections. This technique has wide use for abdominal explorations too.

Why Veterinary Practices Should Consider Offering Endoscopy to Clients

There are multiple reasons why practices should invest in this area of veterinary medicine. Take the use of antibiotics. As professionals we have a responsibility to reduce antibiotic use in livestock and the wider adoption of endoscopy can help us achieve that, given the lower risk of infection from a technique that only requires multiple incisions.

The surgeon's eye is also no match for the use of multiple high-quality lenses. These lenses allow the area in question to be inspected to assess whether or not the position is recoverable. For that reason, and others, an endoscope is a very cost-effective tool for use in large-animal medicine.

A veterinary surgeon's time is extremely valuable too and endoscopy will almost certainly allow a procedure to be performed in less time than if it is done laparotomically. Let us take the example of correcting an LDA. Vets might estimate that it will take 45 minutes to do this laparotomically but what they do not take into account is the preparation and clean-up time. When including preparation and clean-up time, this gives an accurate estimation of time for this procedure as closer to 75–90 minutes. Correcting an LDA laproscopically will generally take 45 minutes from start to finish. That time-saving benefit is considerable, allowing a surgeon to perform more procedures in a single day.

Accuracy when correcting an LDA also favours endoscopy. The Grymer-Sterner technique is a good technique and can be very successful if cases are selected carefully but, as it is effectively a blind punch, it comes with a risk of accidentally puncturing other organs, such as the liver, a situation that will not become apparent until days later when the cow starts to display symptoms of being unwell.

Let us also consider examination of the thorax. As the thorax operates under a vacuum, punching a hole in it introduces air and with that comes risk. There is therefore a massive health contribution from using an endoscopy procedure instead. Ultrasound cannot scan atmospheric air in the thorax but endoscopy, provided it is done with all the precautions observed, is not affected by air and the thorax can therefore be inspected visually.

Conclusion

In conclusion, bovine endoscopy has been established as a superb diagnostic, therapeutic and prognostic technique in veterinary medicine. Its minimally invasive nature, combined with the ability to visualize internal organs clearly and perform corrections, has revolutionized the field of bovine medicine and surgery.

With ongoing developments in endoscopic instruments, techniques and computer imaging, bovine endoscopy continues to facilitate more accurate diagnoses, significantly improve the therapeutic outcomes and enhance the overall welfare of cattle.

As with every technique we have at our disposal, training is key to using it successfully. Private organizations offer training online or face to face. I have personally trained most of the vets who practise ruminant endoscopy in the UK and have been involved in training in Europe and other parts of the world too.

I hope you find this book a useful and informative guide and that it helps our profession move forward to a position where endoscopy is more commonly used in ruminant medicine.

2 Anatomy

Abstract

This chapter elaborates on the various anatomical features that are of interest during endoscopic operations on the bovine patient and the anatomical considerations in the abdomen while exploring sick patients to correct abomasal displacements. The importance of thorax physiology when carrying out explorations while maintaining the negative atmospheric pressure of the pulmonary cavity is discussed.

During endoscopic procedures on the bovine patient, there are many anatomical features that are of interest. This chapter explores these, but firstly let us look at the important anatomical considerations relating to the abdomen while exploring sick patients to correct abomasal displacements. Vacuum is a very important factor in the function of cavities and it explains why the physiology of the thorax is important when carrying out these explorations while, at the same time, maintaining the negative atmospheric pressure of the pulmonary cavity. Let us liken it to a shopping bag of contents kept in a vacuum. In the case of the thorax, that bag contains the lungs, the heart, major blood vessels and pulmonary arteries. For the abdomen, the bag predominantly contains the digestive tract, the kidneys and spleen. As pressure is negative in both cavities, introduction of air, particularly in the thorax can be a complication of endoscopy. This is very rare, but when it does occur it can be serious. Ultrasound is therefore best utilized as a first line of investigation before using endoscopy for any required procedure.

A brief revision of bovine anatomy that is relevant to endoscopy now follows.

Digestive Tract

The purpose of the multiple-compartment digestive system specific to cows and other ruminants is to digest and extract nutrients efficiently from the animal's diet of plant material.

© Sotirios Karvountzis 2023. *Bovine Endoscopy* (Sotirios Karvountzis)
DOI: 10.1079/9781789246681.0002

As ruminants, cows have a unique ability to regurgitate and re-chew partially digested food, a process known as rumination or 'chewing the cud', which enables a further breakdown of tough plant fibres and the extraction of additional nutrients.

Each section of the digestive tract serves a different function:

Mouth and oesophagus: With a large mouth and no upper incisors, cows rely on a hard dental pad on their upper jaw, together with their tongue and lower incisors, to grasp and tear grasses and other forage. This food is mixed with saliva to form a bolus, which passes through the oesophagus after it has been swallowed.

Rumen: The largest and most significant compartment of the bovine digestive system is the rumen, a large fermentation vat capable of holding up to 190 litres (50 gallons) of partially digested food. Billions of microorganisms, including bacteria, protozoa and fungi, are contained in the rumen and through fermentation these assist with the breakdown of cellulose and other complex carbohydrates depending on breed and age of the animal.

Reticulum: The reticulum, a smaller compartment that sits near the front of the rumen, is a storage area for food that has been partially digested and will trap foreign objects that the cow may have ingested, for example metal or stones.

Omasum: The spherical omasum, made up of many layers of tissue with numerous folds which increase its surface area, acts as a filtration system, absorbing water and minerals from the partially digested food. This process will reduce the volume of this material before it passes to the next compartment.

Abomasum: Often referred to as a cow's 'true stomach' because of its similarity to the stomach in monogastric animals, the abomasum secretes hydrochloric acid and digestive enzymes to break down and absorb nutrients.

Small intestine: Most of the nutrient absorption takes place in the small intestine, an organ consisting of the duodenum, jejunum and ileum. Enzymes from the pancreas and bile from the liver help to break down remaining food particles into smaller molecules to be absorbed through the intestinal wall.

Large intestine: In cows, the large intestine is fairly short, and its primary function is to absorb water and electrolytes from indigestible food material.

Rectum: The final section of the digestive system, the rectum, stores faeces before they are expelled through the anus.

Bovine Urinary System

The bovine urinary system, also referred to as the urinary tract, is responsible for the production, storage and elimination of urine. It has several organs and structures that work as a team to maintain the body's fluid balance and eliminate waste products.

This system is vital for maintaining the body's homeostasis by regulating fluid balance and eliminating waste products. If there are abnormalities or disorders in this system, urinary tract infections, urinary stones or other urinary system-related issues can result and these can impact on the health and wellbeing of cattle.

The organs and structures in the bovine urinary system are as follows:

Kidneys: These two bean-shaped organs are located in the bovine's abdominal cavity where they filter waste products, excess water and electrolytes from the blood to produce urine. A secondary function is the regulation of the body's acid-base balance and blood pressure.

Ureters: A muscular tube known as a ureter connects each kidney to the urinary bladder and transports urine from the kidneys to the bladder, propelling it through peristaltic contractions.

Urinary bladder: The urinary bladder, a hollow, muscular organ located in the pelvic cavity of cattle, stores urine until it is eliminated. As it fills with urine it expands and will contract during urination to expel the urine through the urethra.

Urethra: The urethra carries urine from the bladder to the external environment. In bulls, it is longer and more muscular because it also acts as a conduit for semen during reproduction.

Bovine Reproductive System

The reproductive organs and processes found in cattle are known as the bovine reproductive system. This system is responsible for the production, maturation and transport of ova and the subsequent fertilization and development of the foetus.

The reproductive system in cows has a number of main organs:

Ovaries: The two ovaries produce ova (eggs) and release these during the oestrous cycle.

Oviducts (Fallopian tubes): These two small tubes connect the ovaries to the uterus, providing a passageway for ovum to transport toward the uterus.

Uterus: This hollow muscular organ is where fertilization takes place and is where the embryo implants and develops.

Cervix: The cervix, which is part of the uterus, acts as the 'vault door' that protects the uterine interior. It has an essential function in the transport of sperm and also acts as a protective barrier during pregnancy.

Vagina: The vagina is a muscular canal connecting the cervix to the external genitalia. It acts as the birth canal during the delivery of the calf.

Bovine Spleen

The bovine spleen is located in the abdominal cavity of cattle. This important organ, found on the left side of the cow's body, is usually located between the

7th and 8th rib, is dark reddish-brown in colour, has a soft, spongy texture and forms part of the immune system.

Immune function: The spleen helps to filter the blood and remove damaged or old red blood cells, and foreign substances like bacteria and other pathogens too. It also plays a part in producing antibodies and activating immune cells to fight against infections.

Blood storage and release: The spleen acts as a reservoir for red blood cells, releasing these into circulation when necessary. During physical exertion or haemorrhage and at other times of increased demand, the spleen contracts and releases stored red blood cells into the bloodstream to preserve oxygen-carrying capacity.

Haematopoiesis: The spleen has a small role in the production of red and white blood cells in young calves, but as the animal matures this function is taken over by bone marrow and its contribution to haematopoiesis diminishes.

Filtration and destruction: The spleen removes damaged or aged red blood cells and other cellular debris. As well as filtering blood, it also acts as a defence mechanism against infections by helping to destroy bacteria and other pathogens.

Iron metabolism: The spleen stores iron from recycled red blood cells, releasing it for the production of new red blood cells.

Bovine Respiratory System

The bovine respiratory system includes the organs and structures involved in the process of breathing.

Trachea: The trachea, or windpipe, a tube-like structure that connects the larynx to the lungs, is made up of cartilage rings that provide support and prevent collapse. It transports inhaled air down towards the lungs.

Bronchi: The trachea branches into two bronchi, with one leading to each lung. These divide further into smaller tubes, the bronchioles, which continue branching throughout the lungs.

Lungs: Cattle have two large lungs that are positioned in the chest cavity. The bronchioles lead to small air sacs known as alveoli and it is in these that the exchange of oxygen and carbon dioxide occurs. Oxygen from the inhaled air diffuses into the bloodstream, while carbon dioxide, a waste product, moves from the bloodstream into the alveoli to be exhaled. Cattle typically have two lobes in the right lung, three in the left. The right lung contains the cranial lobe and the caudal lobe and the left the cranial lobe, the middle lobe and the caudal lobe. These lobes split further into smaller bronchopulmonary segments, each with its own bronchus and blood supply. The lobes and subdivisions facilitate efficient respiration and gas exchange.

Diaphragm: The diaphragm, a dome-shaped muscular sheet that separates the chest cavity from the abdominal cavity, is essential to the process of breathing, contracting and relaxing and causing the expansion and contraction of the lungs. Located below the lungs and above the abdominal organs, it is composed of a thick, fibrous structure, the central tendon and muscular fibres. During inhalation, the diaphragm contracts, moving caudally and expanding the thoracic cavity, allowing the lungs to fill with air. This contraction creates a negative pressure within the thoracic cavity, drawing air into the lungs. When the animal exhales, the diaphragm relaxes and moves cranially, compressing the lungs and expelling air. As well as its respiratory function, the bovine diaphragm helps to stabilize the abdominal organs and support the spine. It also helps to maintain the position of the digestive organs and has a vital role in maintaining correct posture and balance. Aside from its function in the animal, the bovine diaphragm is used in scientific research and medical training as its anatomical structure and the similarity it has to the human diaphragm make it a useful model for the study of respiratory physiology and surgical techniques.

Bovine Heart

The heart, located in the chest cavity, slightly to the left of the midline, pumps blood throughout the body. Its coordinated contractions enable efficient blood circulation to supply oxygen and nutrients to all parts of the body. Although there are some slight variations in size and structure, its anatomy is similar to that of other mammals, humans included.

Position: In the thoracic cavity, the heart sits between the lungs in a slightly oblique orientation.

Pericardium: The protective sac that encloses the heart, the pericardium is made up of two layers, the outer fibrous pericardium and the inner serous pericardium.

Bovine Mammary Gland and the Teat

The mammary gland, otherwise known as the udder, is responsible for the production and secretion of milk. It is located in the lower abdomen of female cows and is made up of four separate mammary glands, known as quarters, each with its own teat. The quarters are composed of a network of ducts connected to the alveoli, a complex system of milk-producing glands that develop during pregnancy and lactation. The ducts transport milk from the alveoli to the teat, where it can be milked.

The anatomy of the bovine teat is as follows:

Teat cistern: This uppermost part of the teat acts as a reservoir for milk storage before milking.

Teat canal: A small opening located at the tip of the teat, the streak canal is surrounded by a sphincter muscle that helps regulate milk flow. It provides a barrier against pathogens by preventing them from entering the udder.

3 Instruments

Abstract

The instruments used for endoscopic procedures can broadly be classed into three groups: trocars, endoscopes and general surgical equipment. Trocars and endoscopes are very specific to these procedures and very rarely have any uses in other areas of surgery. A number of veterinary surgeons elect to source endoscopes designed for human medicine because this provides a wider and more cost-effective supply.

It would be true to say that the market for the rigid and flexible endoscopes used in farm animal medicine globally is dwarfed by the equivalent for human medicine in Europe and North America alone. Client pressure on farm animal practices to deliver a service that is reasonably cost effective is often a reason why second-hand equipment used in human medicine is considered. Practices should, however, be aware of its disadvantages.

Let us look at the size of human patients, for instance; the tallest patient is likely to be 2.20 m. With this in mind, the biggest rigid endoscope therefore does not need to be longer than 30 cm. In comparison, the biggest candidate for a farm animal procedure can be as heavy as 1 tonne. The length of the endoscope therefore matters. An endoscope that is 45–50 cm long has the advantage of the additional reach required. The additional 15–20 cm, particularly as the patient may be in dorsal recumbency, can give that extra confidence to investigate organs within the abdomen.

As far greater volumes of equipment are manufactured for use in human medicine, sets used in large-animal medicine instantly come with a price premium attached, especially if the equipment is specialized.

This has a direct influence on the second-hand value of farm animal equipment too – as less is manufactured, depreciation is very small compared to equipment used in human medicine, which is being continually updated. For example, a rigid 45–50 cm endoscope used for bovines might depreciate by

© Sotirios Karvountzis 2023. *Bovine Endoscopy* (Sotirios Karvountzis)
DOI: 10.1079/9781789246681.0003

10–30%, but for human medicine a price drop between new and second-hand is likely to be as much as 60–66%. The second-hand market for farm endoscopes is therefore competitive.

When farm endoscopes come at a price premium even for second-hand, practices may question why they want to get involved. The reasons for doing so are many, but let us look at one of these: it sets the practice apart from the competition, it gives it the edge, allowing for better visualization by the surgeon and quicker recovery time by the patient. For many practices it therefore makes financial sense to invest and, moreover, to pay the full amount for the correct equipment; they consider the sets to be a sound investment.

However, if cost is an over-riding consideration, and very often it is, human-medicine equipment is perfectly adequate for use in field conditions, although, as previously discussed, it will have limitations owing to the length of the endoscopes.

The instruments and other equipment you'll require for endoscopic procedures are shown in Fig. 3.1.

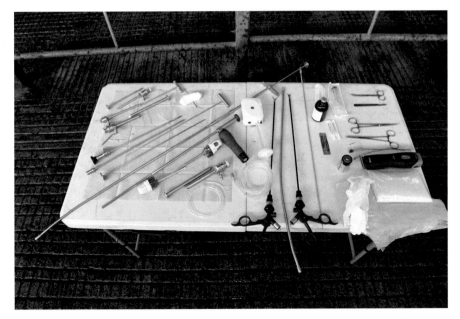

Fig. 3.1. Instruments and other items required for endoscopic procedures in the field. (Copyright Dr Sotirios Karvountzis.)

Instruments required for this technique can be grouped into general surgical instruments, consumables, instruments for the optical portal and working instruments for the working portal. The general surgical instruments we shall need are: a pair of scissors, scalpel blade number 21 to 24, scalpel blade holder, three artery forceps, semi-curved cutting needle and cordless clippers with a number 10 cutting head.

The consumables required are 30-ml syringes, 5-ml syringes, 16-gauge 1.5-inch needles, a 1.5–2-m piece of EP5 nylon suture, a rolled cotton bandage and a disposable drape.

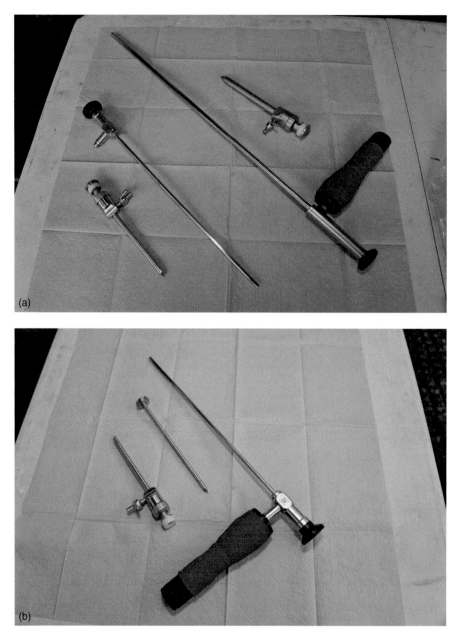

Fig. 3.2. Instruments used to set up the optical portal, including a 5-mm and an 8- or 10-mm rigid endoscope. (Copyright Dr Sotirios Karvountzis.)

The optical portal instruments required are: a 5-mm magnetic trocar, an 8-mm magnetic trocar for the 8-mm rigid endoscope or a 10-mm magnetic trocar for the 10-mm rigid endoscope and a portable light source that operates with rechargeable batteries. All these are shown in Fig. 3.2.

Fig. 3.3. Holding the 5-mm magnetic trocar in preparation for setting up the optical portal. (Copyright Dr Sotirios Karvountzis.)

The 5- and 10-mm magnetic trocars need to have their gas valve set to the off position when they are inserted into the area of interest. This is to prevent inadvertent insufflation because the introduction of air may alter the position of the organs. A typical way of holding these trocars is shown in Fig. 3.3.

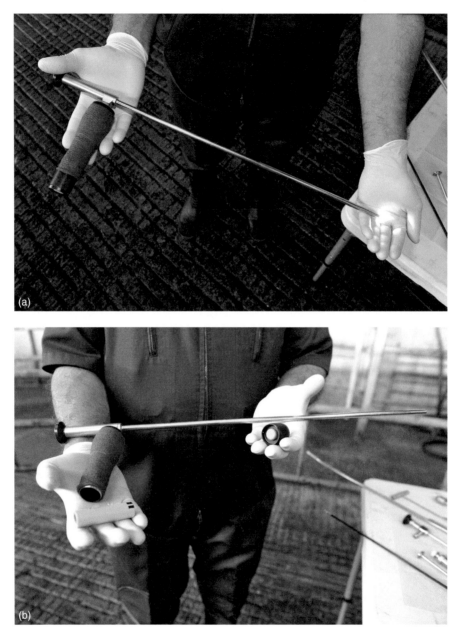

Fig. 3.4. The rechargeable batteries installed in the portable light source used for rigid endoscopes. (Copyright Dr Sotirios Karvountzis.)

The portable light source is a practical solution for the demands of the field. It operates with a rechargeable battery, as shown in Fig. 3.4, that, depending on the make and battery quality, gives the surgeon between 1 and 2 hours of available illumination. Spare batteries can provide longer illumination, with a battery charger installed in the surgeon's vehicle, if required.

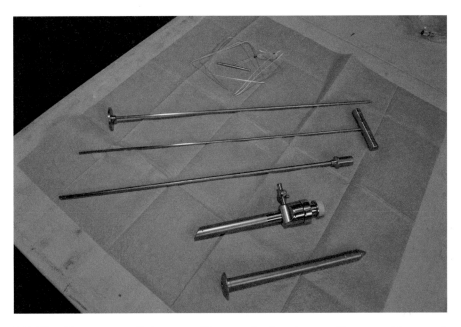

Fig. 3.5. The 13-mm magnetic trocar and its cannula, the abomasal trocar and its cannula, the toggle introducer and, finally, the T-bar toggle. (Copyright Dr Sotirios Karvountzis.)

The working portal instruments required are: a 13-mm magnetic trocar, an abomasal trocar, a toggle introducer, a transfixer, crocodile forceps and endoscopic T-bar toggles containing a pair of nylon sutures. The instruments used to carry out the abomasocentesis are shown in Fig. 3.5.

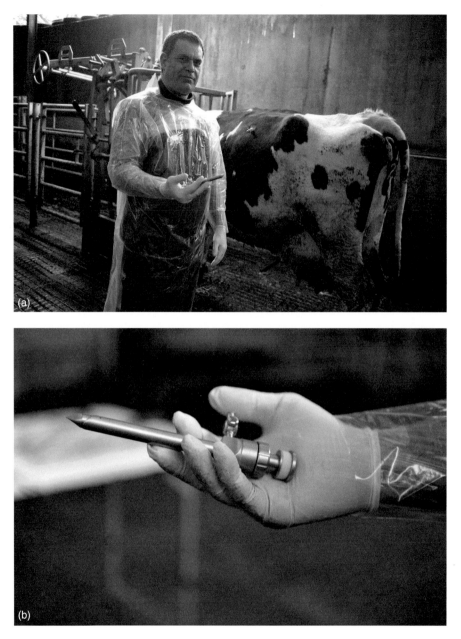

Fig. 3.6. Holding the 13-mm magnetic trocar in preparation for setting up the working portal. Please note that, in the second image, the valve of the gas vent is set up in the 'on' position for illustrative purposes. (Copyright Dr Sotirios Karvountzis.)

The 13-mm magnetic trocar also needs to have its gas vent valve at the off position when it is being set up, to avoid any unintended introduction of air into the abdomen. This is shown in Fig. 3.6.

Fig. 3.7. The abomasal trocar ready to be inserted through the working portal for the abomasocentesis. (Copyright Dr Sotirios Karvountzis.)

The abomasal trocar is a non-magnetic instrument, consisting of the abomasal awl and the abomasal cannula. It is usually 5 mm thick and can be up to 50 cm long. There is no gas vent attached to it, as shown in Fig. 3.7.

Fig. 3.8. The transfixer with its four constituent parts. The longer side of the handle corresponds to the curved part of the cannula's distal end. (Copyright Dr Sotirios Karvountzis.)

The transfixer consists of four parts, two for its needle and two for its cannula, and these are shown in Fig. 3.8. Note the handle of the cannula has two parts of unequal size; the longer side of the handle corresponds to the curved part of the distal end of the cannula.

Fig. 3.9. The 13-mm trocar is the only available trocar in the endoscopic kit that the transfixer will fit through. (Copyright Dr Sotirios Karvountzis.)

Finally, the curve on the distal end of the transfixer is such that it needs a 13-mm trocar. This is shown for illustrative purposes in Fig. 3.9.

Fig. 3.10. Two transfixers with different curvatures at their distal end placed side-by-side for comparison. The transfixer on the left side is used for abomasopexy near the xiphoid process, whereas the transfixer on the right side is used for abomasopexy near the navel. (Copyright Dr Sotirios Karvountzis.)

Finally, note the larger curve at the distal end of the transfixer, when the fixing point of the abomasopexy lies nearer the navel. For comparison, the two transfixers with different curvatures are shown in Fig. 3.10.

Crocodile action forceps are a useful addition to the endoscopic kit when a grasping or cutting action is required. Examples include where the T-bar toggles need to be recovered, peritoneal adhesions broken down or small tangible foreign bodies removed, just to mention a few. Fig. 3.11 shows this instrument.

Fig. 3.11. Endoscopic crocodile forceps are a useful addition to the endoscopic kit. (Copyright Dr Sotirios Karvountzis.)

The end of these forceps can be cutting or grasping. Grasping ends are shown in Fig. 3.12.

Fig. 3.12. Endoscopic crocodile forceps with grasping ends. (Copyright Dr Sotirios Karvountzis.)

Fig. 3.13. Endoscopic crocodile forceps with bending ends. (Copyright Dr Sotirios Karvountzis.)

Another useful feature of this instrument, depending on the model, is having bending ends, when flexibility in the area of interest is required, as shown in Fig. 3.13. Finally, most models have rotating ends, adding another layer of flexibility.

Fig. 3.14. The surgeon observes the insertion of the endoscopic crocodile forceps through the working portal on a standing patient. (Copyright Dr Sotirios Karvountzis.)

The use of the endoscopic crocodile forceps is shown in Fig. 3.14.

The equipment needed for the procedure is summarized below.
For the optical portal:

* 5-mm magnetic trocar and cannula;
* 8- or 10-mm magnetic trocar and cannula;
* rigid endoscope to fit the 8- or 10-mm cannula; and
* portable light source with rechargeable batteries that screws on or clips on the light source end of the endoscope.

For the working portal:

* 13-mm magnetic trocar and cannula;
* abomasal trocar and cannula;
* toggle introducer;
* T-bar toggle;
* transfixer;
* crocodile forceps; and
* nylon suture between 1 and 1.5 m long.

General surgical equipment:

* 1 pair of scissors, curved or straight with curved ends;
* 3 pairs of artery forceps;
* scalpel blade holder;
* wired or wireless (battery-operated) insufflation pump with silicone tubing, filter and 5-mm attachment;
* injectable local anaesthetic;
* 30-ml syringe;
* 5-ml syringe;
* 16-gauge 1.5-inch needle;
* 21–24-gauge scalpel blade;
* one 10-cm rolled cotton mesh bandage; and
* cordless rechargeable clippers with a size 10 clipper blade.

Protective clothing:

* disposable gown;
* disposable arm gloves with shoulder protection; and
* disposable hand gloves.

Items for the care of the endoscopic kit:

* 5-mm long flexible brush;
* 5-mm short flexible brush; and
* 10-mm short flexible brush.

The author's preferred protective clothing for carrying out an endoscopic procedure in the field is a disposable gown, disposable arm-length gloves with shoulder protection and disposable hand gloves, as shown in Fig. 3.15.

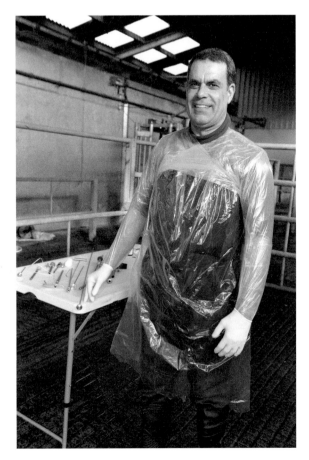

Fig. 3.15. Protective clothing recommended for the procedure. (Copyright Dr Sotirios Karvountzis.)

Fig. 3.16. Recommended length for the extra piece of nylon suture used in abomasopexy. (Copyright Dr Sotirios Karvountzis.)

Carrying out endoscopic procedures in an environment such as the farm, it is near impossible to achieve absolute sterility, but a high level of cleanliness is possible. Using such disposable items allows the surgeon to achieve satisfactory levels of cleanliness and sterility quite quickly.

The author recommends an additional piece of synthetic suture, such as Supramid White USP 6 (EP 8). This extra piece of synthetic suture should be long enough to allow the surgeon to migrate the sutures from the point of abomasopexy to the working portal without them becoming taut or rupturing inside the abdomen. The length should be between 1.5 and 2 m to account for most sizes of cattle patients, as shown in Fig. 3.16.

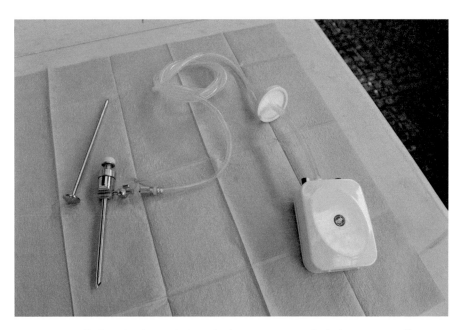

Fig. 3.17. An insufflation circuit consisting of a battery-operated wireless pump, silicone tubing, filter, 5-mm connector and 5-mm magnetic trocar and cannula. (Copyright Dr Sotirios Karvountzis.)

The insufflation pump is a very useful tool when the passive insufflation of the abdomen is not sufficient. A cordless pump, in particular, can sit on top of the patient's back while insufflation takes place. The 5-, 8- or 10- and 13-mm magnetic cannulas are designed in such a way to create a one-way flow of air, from the pump into the cavity we intend to insufflate. An insufflation circuit with a 5-mm trocar and cannula is shown in Fig. 3.17.

Each magnetic trocar consists of two parts: the awl, which may be solid or hollow, and the cannula. The cannula in turn consists of three parts: the seal, the gas vent and the sheath or stem. A 13-mm trocar connected to an insufflation pump is shown in Fig. 3.18.

Fig. 3.18. A 13-mm magnetic trocar connected to an insufflation pump. (Copyright Dr Sotirios Karvountzis.)

The awls we use in bovine endoscopy are all sharp ended and, provided their tips are protected between uses, they need sharpening once every other year. The valve screws onto the stem and contains a magnetic circular diaphragm that controls the flow of air during insufflation and aspiration, as shown in Fig. 3.19.

Fig. 3.19. The valve of a 13-mm magnetic trocar with the awl protruding through and displacing the magnetic diaphragm. (Copyright Dr Sotirios Karvountzis.)

Fig. 3.20. The gas vent of a 13-mm magnetic trocar connected to the silicone pipe of an insufflation pump. In the lacuna of the sheath, we can see where the lacuna of the vent attaches and allows the uni-directional flow of gases during insufflation or aspiration. (Copyright Dr Sotirios Karvountzis.)

The gas vent is where the connector of the silicone pipe from the insufflation pump connects to the trocar. The vent contains a tap held in place by a nut, as shown in Fig 3.20.

Fig. 3.21. The sheath or stem of a 13-mm magnetic trocar with the awl in place, also showing the holes at the distal end permitting the flow of gases while *in situ* and while an instrument is in place. (Copyright Dr Sotirios Karvountzis.)

Finally, the steam or sheath once *in situ* allows the various instruments to be put in place. The diameter of the stem is slightly bigger than that of the instrument it is designed to contain. Furthermore, there are a number of holes at the distal end of the sheath to permit the flow of gases while an instrument is in place, as shown in Fig. 3.21.

For cleaning the endoscopic kit, before it is disinfected, we require the following items:

- 5-mm long flexible brush;
- 5-mm short flexible brush;
- 10-mm short flexible brush; and
- the instrument carrier case that the endoscopic kit is transported in.

The instrument carrier case presents an ideal object in which to disinfect the kit in the field between procedures or between visits. Open the instrument case fully and rest it on a flat surface. Remove the foam mould that the instruments are secured in while the case is in transit. Fill each half of the case with clean water and start cleaning the instruments used externally and when applicable internally in order to remove any visible impurities.

We place any instruments in the water bath and ensure we dismantle any instruments to their constituent parts. For example, for the magnetic cannulas, we unscrew the magnetic valve and remove the circular diaphragm. We also place the remaining stem with the gas vent attached in the water bath, with the tap left in the on position.

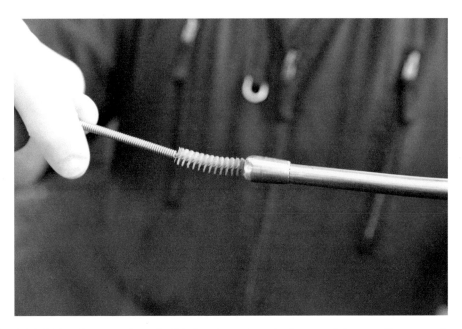

Fig. 3.22. Using the 5-mm long flexible brush to clean the lacuna of the transfixer cannula. (Copyright Dr Sotirios Karvountzis.)

Other instruments, such as the transfixer, we dismantle to the two cannula parts and its two needle parts and then place them in the case for cleaning. Also remove any plastic or rubber washers from the trocars and place them in the solution for cleaning.

Use the long 5-mm flexible brush to clean the lacuna of the long transfixer cannula, as well as the abomasal cannula. Start from one end and then repeat the process from the other end, as shown in Fig. 3.22.

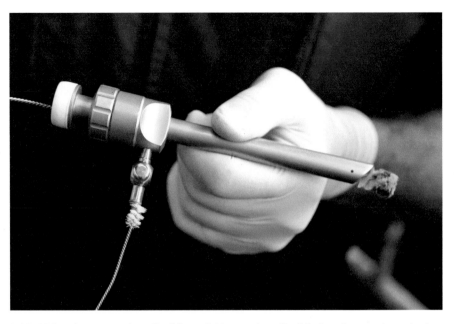

Fig. 3.23. Using the 5-mm short flexible and 10-mm short flexible brushes to clean the lacunas of the cannula stem and the gas vent, respectively. (Copyright Dr Sotirios Karvountzis.)

Use the 5-mm short and 10-mm short flexible brushes to clean the lacuna of the magnetic trocars. We use the 10-mm short flexible brush to clean the valve and stem of the cannula, and we use the 5-mm short flexible brush to clean the gas vent, both shown in Fig. 3.23.

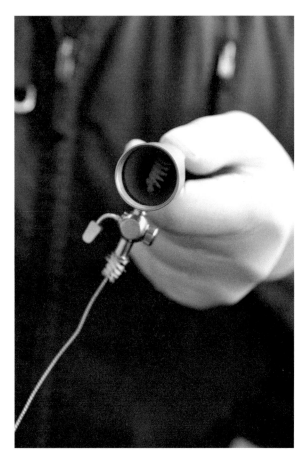

Fig. 3.24. Using the 5-mm short flexible brush to clean the lacuna of the gas vent while the tap of the gas vent is in the open position, so that the brush reaches the point where it joins the stem lacuna. (Copyright Dr Sotirios Karvountzis.)

Please note that while cleaning the gas vent the tap needs to be in the open position, so that the end of the brush will clean the whole of the gas vent lacuna, all the way to the point it joins the stem lacuna of the trocar cannula, as shown in Fig. 3.24.

There are three types of disinfection of the endoscopic kit:

- autoclave sterilization;
- cold sterilization; and
- ethylene oxide sterilization.

All endoscopic instruments are suitable for autoclave sterilization, except some endoscopes and their rubber washers. The surgeon or nurses need to check first whether the endoscopes in particular are suitable for autoclaving.

Disinfection of the surgical kit is an utter necessity, but it represents a downtime, during which the surgical kit is unavailable. Cold sterilization is an excellent means of delivering disinfection in the field, allowing the surgical team to continue their working day with minimal interruptions. To carry out cold sterilization, fill each half of the case with clean water and add the approved disinfectant for endoscopic sterilization in one of the two halves. Depending on the dilution of the disinfectant, leave the instruments for the prescribed standing time. The dilution and standing time differ with each disinfectant product.

Ethylene oxide sterilization is very common in veterinary surgery, because it allows gentle handling of delicate instruments, such as fibre-optic endoscopes or rubber washers of instruments. An ethylene oxide chamber would be based at the practice premises; therefore the endoscopic kit needs to be returned to base for disinfection at regular intervals. There are other serious considerations about this method because it is a dangerous substance, extremely flammable and under certain conditions explosive.

The 5-mm or the 8- or 10-mm rigid endoscopes are inserted through the cannula of the 5-mm or 10-mm magnetic trocars, respectively, in order to complete the set-up of the optical portal, as shown in Fig. 3.25.

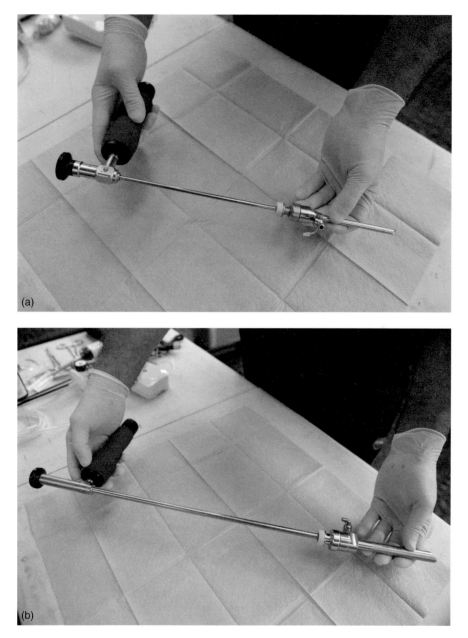

Fig. 3.25. Depending on the size of the rigid endoscope used for the procedure, the relevant size cannula is required to set up the optical portal. (Copyright Dr Sotirios Karvountzis.)

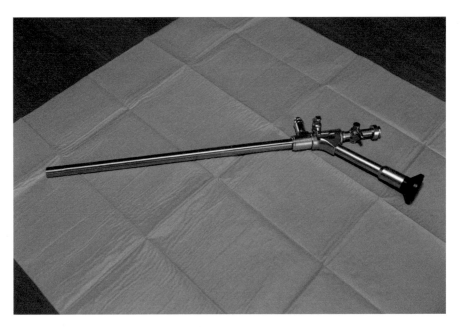

Fig. 3.26. A rigid 33-cm-long and 10-mm-wide combined endoscope that contains the optical section, which attaches to the instrument at a 30-degree angle. The working section is hollow and forms part of the main body of the instrument. The light source attachment bay is built perpendicularly to the main body of the instrument, at the opposite site to where the optical section attaches. Finally, two insufflation gas vents complete the structure. (Copyright Dr Sotirios Karvountzis.)

Combined endoscopes are gaining popularity in farm animal veterinary medicine. They come with considerable advantages, particularly in minimizing intrusion to the patient for exploratory and corrective techniques. A combined endoscope is shown in Fig. 3.26.

Fig. 3.27. A rigid 10-mm combined endoscope that shows details of its two insufflation gas vents to allow insufflation and aspiration of gases and fluids. (Copyright Dr Sotirios Karvountzis.)

A combined endoscope comes in a variety of forms; a commonly used one comprises the optical section and the working section. The optical section includes the fibre-optic lens, which is located in the main stem of the instruments and runs parallel and adjacent to the working section. The optical section culminates in the eye piece. The eye piece is usually connected at a 30-degree angle to the main stem, so as not to interfere with the function of the working channel. Finally, the light source attaches to the optical channel, so that it permits the illumination of the examination area. The working section runs parallel and next to the optical fibre-optic channel and comprises the lumen, through which the instruments used in the working portal are inserted. Because combined endoscopic instruments need to fulfil two purposes, they tend to be bulky; 10 mm in diameter is a very common size for combined endoscopes. Finally, to facilitate the aspiration or insufflation of gas or fluids, one or more gas vents are included with this type of instrument. The gas vents of a combined endoscope are shown in Fig. 3.27.

Fig. 3.28. A rigid 10-mm combined endoscope, showing the components of the examination end of the endoscope. In it we can see the working channel lacuna and the optical segment. The latter includes the fibre-optic end and the light source outlet. (Copyright Dr Sotirios Karvountzis.)

A further detail of the combined endoscope is shown in Fig. 3.28. The examination end of this instrument contains two sections, the viewing end of the optical segment and the lacuna of the working part. The optical part includes the viewing fibre-optic and the light source channel. The light source channel is normally a circular-shaped channel that sits just inside the perimeter of the instrument and surrounds all the aforementioned components. The lacuna is big enough to accommodate a 5-mm instrument (other images will show the crocodile forceps). The total width of this endoscope is 10 mm.

Fig. 3.29. A portable light source, attached to a rigid endoscope, before it is used in the field. (Copyright Dr Sotirios Karvountzis.)

Combined endoscopes are designed to have portable or wired light sources attached. A portable light source that operates on disposable or rechargeable batteries is an ideal choice for use in the field. These light sources are easy to carry, are prone to very little damage and are, in most cases, water resistant. Depending on their manufacturer, they can have a clip-on or screw-on attachment to the light source dock of the endoscope. One such light source, attached to a combined endoscope, is shown in Fig. 3.29.

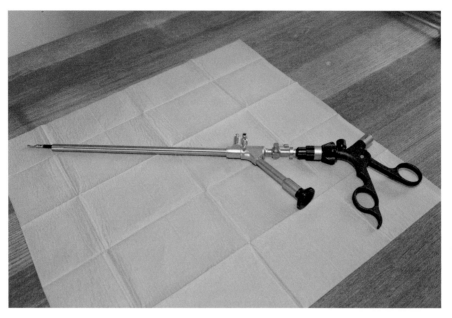

Fig. 3.30. A combined endoscope used with crocodile forceps that bear a cutting scissor tip. (Copyright Dr Sotirios Karvountzis.)

The versatility of combined endoscopes allows the endoscopic surgeon to use it for a multitude of purposes. From diagnostic biopsy to corrective removal of adhesions to removal of foreign bodies, if their size permits. A 5-mm set of crocodile forceps is often used though the working lacuna of the endoscope, with these forceps usually having interchangeable ends, such as a gripping jaw to cutting scissors. An example of crocodile forceps used in the field with a combined endoscope is shown in Fig. 3.30.

To prepare the combined endoscope in the field, a 10-mm magnetic trocar is used to set up the portal that the endoscopic instrument will be inserted through. It is worth noting that the single portal in this instance is a combined optical and working one. Through the combined endoscope a choice of working instruments can be used, one at a time, with the most popular choice being the crocodile forceps with interchangeable tips. Finally, the light source is attached to the endoscope. This set-up is shown in Figs 3.31, 3.32, 3.33 and 3.34.

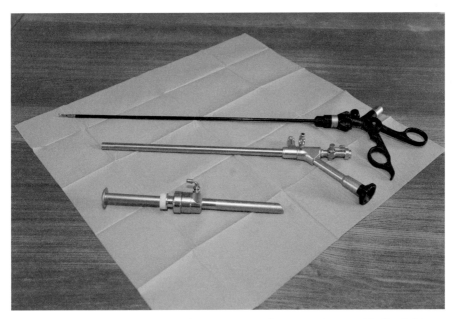

Fig. 3.31. A combined endoscope with instruments and accessories used in the field, showing the 10-mm trocar and crocodile forceps with scissor tip. (Copyright Dr Sotirios Karvountzis.)

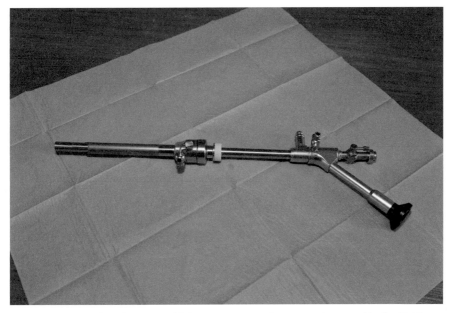

Fig. 3.32. A combined endoscope with instruments and accessories used in the field, inserted through the 10-mm cannula that is used to set up the working portal. (Copyright Dr Sotirios Karvountzis.)

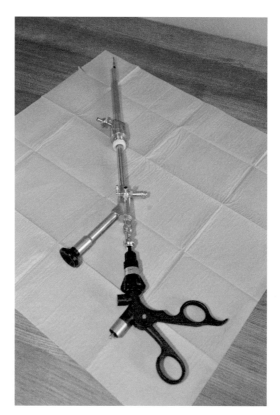

Fig. 3.33. A combined endoscope with instruments and accessories used in the field, showing how the 10-mm cannula and crocodile forceps are set up. (Copyright Dr Sotirios Karvountzis.)

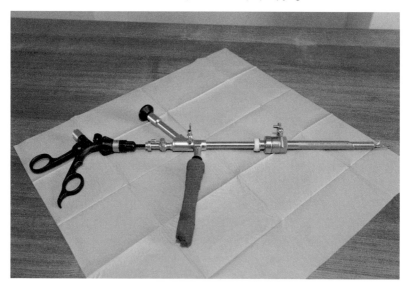

Fig. 3.34. A combined endoscope with instruments and accessories used in the field, showing how the 10-mm cannula, crocodile forceps and portable light source are set up. (Copyright Dr Sotirios Karvountzis.)

Fig. 3.35. A 10-mm combined endoscope, 8-mm endoscope, 5-mm endoscope, 3-mm endoscope, a portable LED light source and the photographic instruments used for recording endoscopic operations in the field. The cameras used for recording throughout this publication are a Panasonic Lumix DC-GH5 digital camera and a KoPa MC500W-G2 WiFi digital camera. (Copyright Dr Sotirios Karvountzis.)

There are a number of devices suitable for use in the field which meet the increasing need to record and share images and videos of endoscopic work. Some of those photographic devices are shown in Fig. 3.35.

The portable LED light source shown in Fig. 3.36 operates with two AA batteries and produces approximately 200 lux of illumination. These light sources are easy to carry on the farm and are water resistant, although wrapping them with a protective bandage protects them from contamination by water or other impurities. These light sources either clip or screw on to the light dock of the endoscope in use.

Fig. 3.36. A portable LED light source showing the illuminating end that docks to the endoscope in use. This particular model has a screw-on fitting. (Copyright Dr Sotirios Karvountzis.)

The Panasonic Lumix DC-GH5 digital camera used in this publication and the way it attaches to the various endoscopes is shown in Fig. 3.37. This camera has a resolution of 5184 by 3888 pixels and is relatively heavy, weighing approximately 710 g. During use in the field, it needs to be protected from water contamination and dirt.

(a)

(b)

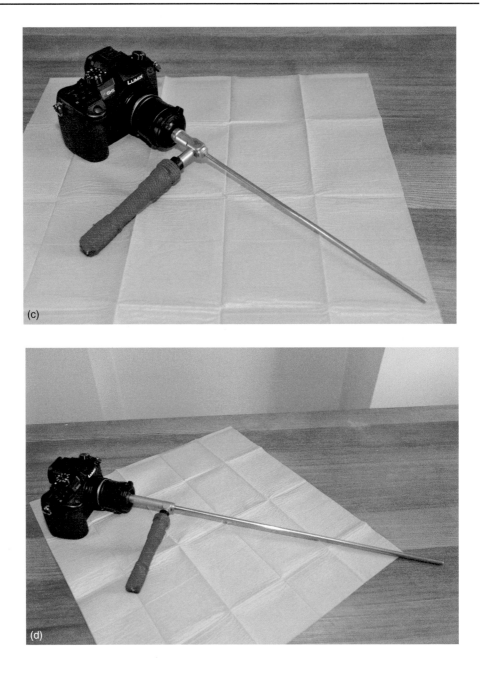

(e)

Fig. 3.37. The fittings and attachments of the Panasonic Lumix DC-GH5 digital camera are shown here. Figs 3.37a and b show the fitting and its protective cap for attachment to the optical end of the endoscope. Figs 3.37c, d and e show the camera and how it attaches to various endoscopes used in this publication. Note the comparative size between this digital camera and the 3-mm endoscope. (Copyright Dr Sotirios Karvountzis.)

The KoPa MC500W-G2 WiFi digital camera used in this publication and the way it attaches to the various endoscopes is shown in Fig. 3.38. This camera has a resolution of 2592 by 1944 pixels and is of low weight, weighing approximately 150 g. During use in the field, it also needs to be protected from water contamination and dirt, including its antenna, which should not be obstructed. This camera allows viewing of the findings on a mobile phone, connecting the camera to the phone through its application program, which in turn allows internal recording of still images and videos of the procedure.

(a)

(b)

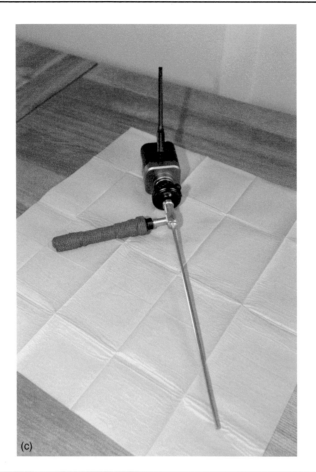

(c)

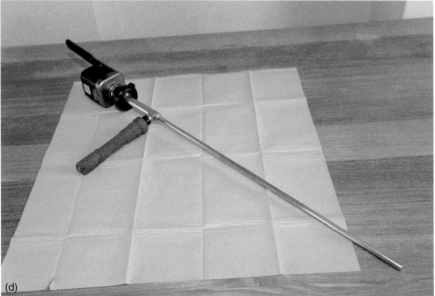

(d)

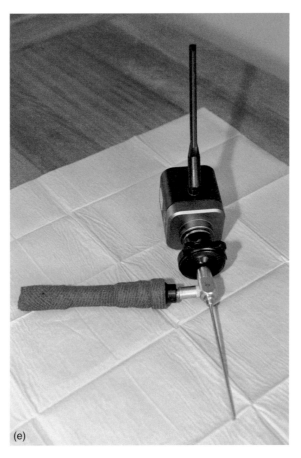

Fig. 3.38. The fittings and attachments of the KoPa MC500W-G2 WiFi digital camera are shown here. Figs 3.38a and b show the fitting and its protective cap for attachment to the optical end of the endoscope. Figs 3.38c, d and e show the camera and how it attaches to various endoscopes used in this publication. (Copyright Dr Sotirios Karvountzis.)

4

Preparation and Patient Restraint

Abstract

As in most surgical procedures, surgeon and surgical field preparation is key to the outcome of the technique, but brief planning on where to restrain the patient before the commencement of the procedure can prevent a number of setbacks during the operation. This can guarantee a smooth flow of the technique, a successfully treated patient and a satisfied client.

General considerations for the procedure begin with patient restraint. When appropriate facilities are not present, a patient can be restrained with just a head halter, provided this can be tied firmly and securely in a clean corner. Alternatively, an examination crush (or handling chute) can be utilized. These installations are designed to restrain the most fractious animals, with lots of horizontal bars that would interfere with the handling of many endoscopic instruments. The surgeon may consider a configuration to restrain the animal, as in Fig. 4.1, in which there is adequate head restraint and prompt release in case the standing patient becomes unexpectedly recumbent.

Fig. 4.1. Suggested head restraint for the patient, before the commencement of the endoscopic operation. (Copyright Dr Sotirios Karvountzis.)

Finally, for the particularly unsettled patient, leg shackles can be utilized, as shown in Fig. 4.2. These may prevent the surgeon from being kicked, but they may equally generate further distress in the patient, just by their mere presence. If the surgeon decides to use them, the author advises they are placed wide apart so that they do not interfere with the animal's stance.

Fig. 4.2. The use of leg shackles is an option to prevent the surgeon from being kicked. (Copyright Dr Sotirios Karvountzis.)

Fig. 4.3. The choice of site for the optical portal on the left paralumbar fossa for a left displacement of abomasum correction. (Copyright Dr Sotirios Karvountzis.)

The selection of the site of the optical portal on the left flank for a standing patient can vary and it depends on the purpose of the procedure. The site of the optical portal for a left displacement of abomasum in the left flank is shown in Fig. 4.3. In this image, the optical portal is located approximately a hand's width ventrally from the transverse processes and a hand's width caudally from the last rib.

Other sites for the optical portal on the left flank include a site caudally to the aforementioned one that lies a hand's width ventrally from the transverse processes and a hand's width cranially from the tuber coxae. This optical portal site is ideal when exploration of the retroperitoneal space is required. Also, it can be utilized when correcting a left displacement of the abomasum and there is concern about the presence of a large-sized rumen or a large-sized displaced abomasum. Choosing this site that is located caudally on the paralumbar fossa is less likely to unintentionally traumatize enlarged viscera. Also, when choosing this site as the optical portal, the surgeon needs to bear in mind its limitations when observing the cranial abdomen from it, as well as the limitation that the neighbouring tuber coxae imposes on the range of movement of the endoscope.

The preparation of the surgical site is minimal, requiring depilation using cordless clippers with a number 10 or number 40 clip head, as shown in Fig. 4.4.

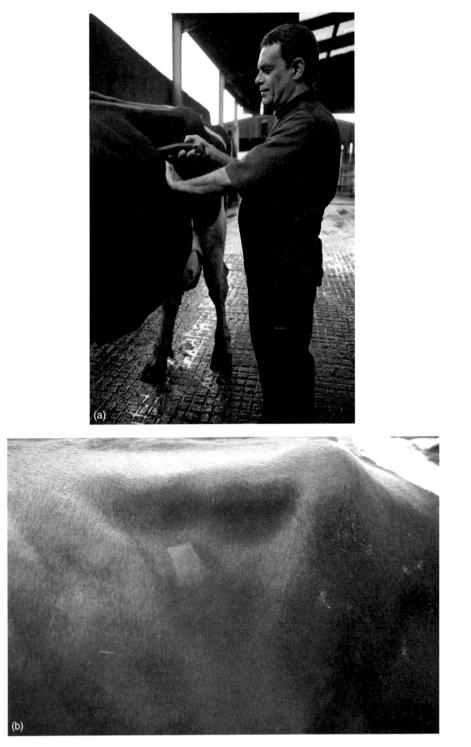

Fig. 4.4. (a) Depilation of the optical portal site on the left paralumbar fossa for the standing patient. (b) The clipped area required is minimal. (Copyright Dr Sotirios Karvountzis.)

Fig. 4.5. Ensure the neighbouring tuber coxae is also depilated to prevent debris from being introduced with the endoscope into the abdomen. (Copyright Dr Sotirios Karvountzis.)

To complete the depilation of the optical site of the left flank, we need to include the tuber coxae as well, as shown in Fig. 4.5. During endoscopy, the endoscope will rest or touch on the tuber coxae, particularly when viewing cranially in the abdomen. Removing the hair from this anatomical feature will prevent loose items from being dragged into the abdomen with the moving endoscope.

Fig. 4.6. Location of the optical portal on the right side of the standing patient. (Copyright Dr Sotirios Karvountzis.)

The selection of the optical portal on the right side of the standing patient mirrors the one of the optical portal on the left side. Its location is shown in Fig. 4.6.

Ventrally, optical and working portals are required for exploratory and corrective techniques. To access those, the patient needs to be placed in dorsal recumbency. Consideration should be given to the equipment and workforce required before the operation commences. The location of the ventral optical portal can be a hand's width caudally from the xiphoid process and a hand's width abaxially from the linea alba on the right-hand side of the patient, as shown in Fig. 4.7.

Fig. 4.7. The location of the ventral optical portal on the right-hand side of the patient. (Copyright Dr Sotirios Karvountzis.)

The ventral optical portal can also be set up a hand's width cranially from the navel and a hand's width abaxially from the linea alba on the left-hand side of the patient, as shown in Fig. 4.8.

Fig. 4.8. The location of the ventral optical portal on the left-hand side of the patient. (Copyright Dr Sotirios Karvountzis.)

Regarding local anaesthesia of the operating area, the author's preferred regime is local infiltration with 15–20 ml of 50 mg/ml procaine hydrochloride that also contains 0.02 mg/ml epinephrine. Equally good results can be achieved with local infiltration of 15–20 ml of 2% lignocaine hydrochloride that also contains 0.02‰ epinephrine.

The local anaesthetic needs to be deposited centrally in order to achieve nerve block of the nerves responsible for the portal incision area. Depending on the site of the portal, the relevant nerves travel in a dorso-cranial to ventro-caudal manner. The ideal point of placing and fanning the nerve block is shown in Fig. 4.9.

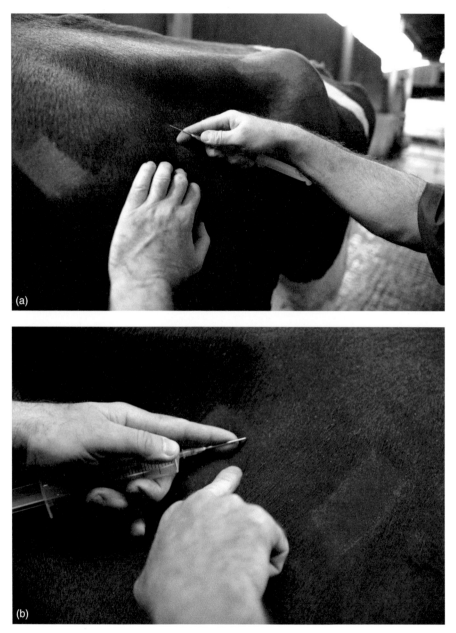

Fig. 4.9. Bearing in mind that for most portals the nerves we need to block run in a cranio-dorsal to ventro-caudal route, we need to place and fan the nerve block at the top left corner of the clipped area for portals on the left side of the body (a) and the top right corner of the clipped area for portals on the right side of the body (b). (Copyright Dr Sotirios Karvountzis.)

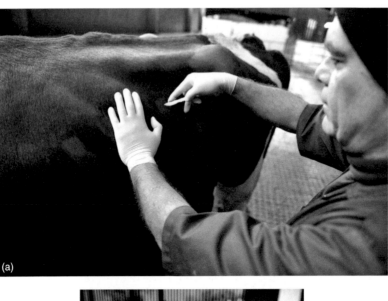

Fig. 4.10. The preferred way of holding the scalpel while incising through the skin. (Copyright Dr Sotirios Karvountzis.)

The skin incision, in preparing the setting up of the portals, should not cut through the whole of the abdominal or thoracic wall, but just through the skin. Should it include the whole of the wall, uncontrolled insufflation of the cavity may occur. The author's preferred way of holding the scalpel is shown in Fig. 4.10a,b. To achieve this, the skin can be incised in a single

Fig. 4.11. The location of the working portal for a small- to medium-sized cow, that being at the last (12th) intercostal space. Abomasopexy will take place near the xiphoid process. (Copyright Dr Sotirios Karvountzis.)

stabbing shallow motion that would avoid penetrating through the whole of the wall. If the incision is found to be insufficiently long and needs to be enlarged, this can then be extended in a controlled manner.

While the optical portal is used for passive tasks, such as inspections that aid the establishment of a diagnosis, the working portal is required mostly for corrections. We need a suitably located working portal when correcting a left- or right-displaced abomasum, removing a foreign body or obtaining a tissue biopsy, to mention a few.

The location of the working portal depends on a number of factors. For the standing patient, the location of the working portal depends on whether we need to work on the left or the right side of the body. A further factor is the size of the patient: for small- to medium-sized cows the working portal is located at the 12th (last) intercostal space, as shown in Fig. 4.11, whereas for a large-sized cow (approximately 900 kg) the working portal can be located at the 11th (penultimate) intercostal space, as shown in Fig. 4.12. As the size of the cow increases, so does the need to access the cranial abdomen and parts of the displaced abomasum. The surgeon should exercise caution when selecting the penultimate intercostal space in a small cow, as this can be the thoracoscopic site for the optical portal. In such a case, the risks associated with thoracoscopy should be taken into account. A further factor is whether the

Fig. 4.12. The location of the working portal for a large-sized cow, that being at the penultimate (11th) intercostal space. Abomasopexy will take place near the xiphoid process. (Copyright Dr Sotirios Karvountzis.)

abomasopexy will take place near the xiphoid process or near the navel. In the former, the working portals are located in the intercostal spaces, as described previously. In the latter, the working portals are located in the left paralumbar fossa, halfway down to be more precise, as shown in Fig. 4.13.

The various possible working portals that can be utilized on the left side of the standing patient are depicted in Fig. 4.14.

Fig. 4.13. The location of the working portal for a cow, that being halfway down the paralumbar fossa. Abomasopexy will take place near the navel. (Copyright Dr Sotirios Karvountzis.)

Fig. 4.14. The various possible working portals that can be utilized on the left side of the standing patient: (1) working portal, standing technique, xiphoid process abomasopexy, large-sized patient; (2) working portal, standing technique, xiphoid process abomasopexy, small- to medium-sized patient; (3a and 3b) working portal, standing technique, navel abomasopexy; (4) optical portal; (5) tuber coxae. (Copyright Dr Sotirios Karvountzis.)

Fig. 4.15. Location of the working portal for a standing patient on the right-hand side. There are two locations shown here, one on the 12th intercostal space and another on the 11th intercostal space. (Copyright Dr Sotirios Karvountzis.)

Fig. 4.16. Location of the working portal on the right-hand side. Abomasopexy will take place here near the navel. (Copyright Dr Sotirios Karvountzis.)

The locations of working portals on the right side of the standing patient are shown in Fig. 4.15.

The ventral working portal is usually set up a hand's width cranially from the navel and a hand's width abaxially from the linea alba on the right-hand side of the patient, as shown in Fig. 4.16.

5 Exploratory Laparoscopy

Abstract

A considerable number of ailments in cattle originate from the abdomen. Exploratory laparoscopy offers a formidable tool in the large-animal surgeon's diagnostic repertoire. Depending on the location of the portal, the digestive, reproductive or urinary systems can be inspected, and more. With minimal intrusion and disruption to the patient's daily routine, in most cases enough evidence can be collected to establish a prognosis for the disorder affecting the patient.

The General Exploratory Laparoscopic Procedure

Every abdominal endoscopic procedure starts with the same steps as for the left exploratory laparoscopy. Whether we prepare for a two-step endoscopic repositioning of a left displacement of the abomasum (LDA) or for the treatment of ovarian cysts, the start of each of these procedures is an exploration of the abdomen.

During this step, insufflation of the abdomen takes place, either actively or passively. Insufflation is a very important step in any endoscopic procedure as it allows the observation of the target organs with the endoscope. Without a gas bubble surrounding the leading end of the endoscope, the target organs would collapse around the instrument, making observation impossible.

Consideration is given to whether carbon dioxide or atmospheric air should be used for the insufflation. For carbon dioxide, the advantage is that a clean and filtered product is deposited around aseptic tissue. However, there is concern that introducing carbon dioxide into organic tissue may increase the likelihood of lowering the pH of that tissue. Research has proven these concerns to be unfounded. Also, for most endoscopic procedures that are carried out in the field, atmospheric air is readily available.

Insufflation can be passive or active. The intra-abdominal atmospheric pressure is negative, therefore when a portal is established between the abdomen and the environment air tends to flow from the environment into the abdomen. This is a passive passage of air and can be simply achieved by opening the valve of a magnetic cannula, once the cannula is *in situ*. Active insufflation relates to the introduction of air into the abdomen by means of a pump. The pumps used are very often converted oxygen pumps, similar to those operated in fish tanks. Such pumps can be powered by mains electricity or are cordless and battery-operated; the latter is quite convenient to set up in the field. The advantage of passive insufflation is that it is simple and readily available, but its downside is slow speed. Active insufflation can introduce sufficient air within minutes, even for the endoscopic exploration of a patient in dorsal recumbency.

The indication of laparoscopic exploration through the left paralumbar fossa is to observe the dorso-left abdomen, the rumen, the spleen, the left side of the diaphragm, the abomasum that has displaced to the left, the left kidney, the left retroperitoneal space with the uterus, broad ligament, left ovary, urinary bladder and parts of the small and large intestine. The location of the optical portal used for an exploratory laparoscopy lies on the left paralumbar fossa, between half a hand and a whole hand's width caudally to the 13th rib and ventrally equidistant to the left lumbar transverse processes.

Resorting to laparoscopic exploration of the right paralumbar fossa is rare due to the proximity to the omentum and its interference in setting up the optical portal. Once the portal is successfully established, the indication of a right exploratory laparoscopy is to assess the dorso-right abdomen, the liver, the gall-bladder, the abomasum, the small intestine, the large intestine including the caecum when it is displaced, the right kidney and the right retroperitoneal space with the uterus, broad ligament, right ovary and urinary bladder. The location of this optical portal mirrors the location of the optical portal on the left paralumbar fossa.

The two-step laparoscopic exploration offers us the ability to complete the examination of the abdominal cavity and inspect most of the viscera. The indication of a ventral exploratory laparoscopy is to inspect the ventral abdomen, rumen, reticulum, omasum, abomasum, liver, pancreas, omentum, diaphragm, small intestine, large intestine and the ventral retroperitoneal space, with the uterus and urinary bladder. The ventral portals can be located in a number of sites on the abdominal wall, in order to facilitate inspection of the area of interest. They are usually, but not exclusively, found between half a hand and a whole hand's width caudally to the xyphoid process and equidistant from the linea alba, either on the left or the right side.

The primary objective in setting up the optical portal is to control any leakage of atmospheric air in the abdomen, therefore controlling the incidence of pneumoperitoneum. Should the optical portal cannula not fit precisely to the skin incision, air would inadvertently leak into the peritoneum. Such a steady low-flow leak is likely to restore atmospheric pressure unilaterally in the abdomen, on the side of the optical portal. This unequal restoration may put additional pressure and move further ventrally or even unexpectedly reduce without fixing any low-lying left-displaced abomasum. Such considerations are important if we intend to complete a left displacement of abomasum as a one-step endoscopic operation and avoid having to substitute with a two-step procedure.

By using a number 21 scalpel blade attached to a scalpel holder, preferably with graduation in millimetres on its handle, we only incise the skin layer while sparing the remaining abdominal muscle layers. The purpose of selectively incising the abdominal layers is to also prevent pneumoperitoneum. The scalpel holder is held horizontally in the palm of the hand and the skin incision is carried out in a brief and quick stabbing motion. The length of the skin incision should be approximately 1 cm in order to match the largest instrument located in it – the 8- or 10-mm magnetic cannula – depending on the manufacturer of the endoscopic kit. The skin incision is elongated as necessary to achieve the desired length.

During the set-up of the optical portal, the advance of the trocar cannot in any way be visualized and, as such, this constitutes a blind part of the procedure. The risk of inadvertent visceral trauma to organs such as the rumen, abomasum or the spleen needs to be mitigated. The proposed approach is to introduce a smaller magnetic trocar and cannula set first, before the larger set is *in situ*. The introduction of the smaller set serves as the pathfinder to determine the correct angle and direction of advance of the trocar–cannula sets.

The first set to introduce is the 5-mm magnetic trocar and cannula, followed by the 8- or the 10-mm devices, depending on the manufacturer of the endoscopic kit. Before its insertion, the valve of the magnetic trocar and cannula set should be in the 'off' position, to prevent pneumoperitoneum. Then the 'T' part of the trocar is placed across in the palm of the left hand, against the soft base of the thumb. The stem of the trocar and cannula set lies along the palm of the hand, stabilized between the median and paramedian fingers. When ready to insert the first trocar and cannula set, fingers and thumb are flexed, forming a fist that engulfs the set in the surgeon's hand.

The sharp tip of the set is carefully introduced through the skin incision and then advancing through the remaining abdominal wall with a short, sharp stabbing motion, while the direction of insertion is aimed medially and cranio-dorsally. When setting up the optical portal on the left or right paralumbar fossa, the surgeon should hold the magnetic set at a 60-degree angle to the transverse plain, 30 degrees to the frontal plain and 30 degrees or less to the sagittal one. When setting up the optical portal on the ventral abdomen, while the trocar set is aimed medio-cranially, it should be held at a 30-degree angle to the frontal plain. The intention of guiding the set in such direction is to avoid contacting any of the abdominal viscera that are located in the area, either normally or abnormally. Once *in situ*, the trocar is promptly retracted slightly or completely removed in order to prevent its sharp tip scraping or traumatizing viscera.

After the set is in place, the valve is opened and the surgeon examines the airflow. Normally, the surgeon would carry out the following examinations in a single inspection, but for facilitating learning we have broken these down to individual components. Firstly, we need to determine whether there is movement of air through the valve. Should there be movement, a hissing sound is heard, created by the passage of gas through a narrow-bore hole. In the absence of such sound, it would be unlikely for the trocar and cannula set to be in the correct position and it may possibly lie between the abdominal muscle layers. In such an event, the set needs to be removed and repositioned, starting from the first steps as outlined above. Secondly, we need to ascertain whether there is insufflation or aspiration of gases. The easiest way for the surgeon to determine this is by bringing their face in close proximity to the valve of the magnetic cannula. The aspiring gases would create a tingling sensation on the face, whereas insufflated gases would not have this effect. Thirdly, the smell of the exchanged gases needs to be verified. In particular, we are interested in identifying smells similar to ruminal, abomasal or intestinal contents, as this would indicate that we have accidentally entered the rumen, abomasum or intestine, respectively.

As the first trocar and cannula set to be used is the smallest one, which measures 5 mm, in the event of inadvertent entry to the abomasum or another organ, the severity of any complications is likely to be small. As soon as the surgeon establishes the correct angle and direction to insert the magnetic trocar and cannula set, the 5-mm set is removed and this process is repeated step by step with either the 8- or 10-mm magnetic set, depending on the manufacturer of the endoscopic kit. Once the larger magnetic set is in place, then the optical portal is ready and we can now place our endoscope with the attached light source to begin the inspection.

It is worth noting that, due to their small size, the portals are not sutured post-operatively. In my experience, in normal circumstances they heal within 24–48 hours without complications. If the portal skin incisions require closure, wide stainless steel skin staples are recommended.

While exploring, the surgeon is monitoring for changes in the morphology and location of the viscera. Morphological characteristics of interest are changes in colour of the mucosae, changes in their ability to reflect or transmit light and alterations in organ size. Further noticeable characteristics are the presence of haematoma, haemorrhage, discharge including of purulent content, fibrin, adhesions, ulceration, tumours and foreign bodies. Assessing blood extravasation is of particular significance, as it may provide an indication of the duration and severity of the insult. When carrying out endoscopic explorations, the types of haematoma we are interested are petechia (smaller than 2 mm), purpura (3 mm–1 cm) and ecchymosis (larger than 1 cm).

A description follows of the abdominal pathological conditions that can be identified endoscopically.

Peritonitis

Bovine peritonitis, otherwise known as peritonitis syndrome, is characterized by inflammation of the peritoneum, the thin membrane that lines the abdominal cavity and covers the abdominal organs. The condition, which is specific to bovines, is caused by viral, bacterial and fungal infections, traumatic injuries, post-surgery complications and a variety of other factors.

It can be categorized as either primary or secondary peritonitis. Primary peritonitis arises when the peritoneum is directly affected by the infection, while secondary peritonitis occurs as a result of injury or infection in other parts of the body that spreads to the peritoneum. Bacterial infections, particularly those caused by *Escherichia coli* and *Trueperella pyogenes*, which can enter the abdominal cavity through the digestive tract, uterus or via traumatic injuries, are commonly associated with bovine peritonitis.

Diagnosis typically involves a careful physical examination, analysis of clinical signs, and laboratory tests such as blood and abdominal fluid analysis. Ultrasound or radiography can also be used to assess the condition of the abdominal organs.

The following paragraphs were originally published in https://doi.org/10.1136/vetreccr-2020-001135.

I would like to make a special mention in this section of splenoptosis, which I have very often found to be accompanied by peritonitis. Cases of splenoptosis have been extensively described in human medicine. In veterinary literature there is a limited number of references for such splenic anomalies, with ectopic spleen described in dogs and rabbits, but no cases found of acquired partial splenoptosis in cattle.

I have encountered this condition a number of times in cattle. In these cases, it coincided with left displacement of abomasum (LDA). Each time, there was formation of adhesions of the left dorso-cranial abdominal region, which may have led to adhesions forming between the displaced abomasum and the spleen. As the displaced abomasum changes the level of its displacement on the left, this in turn puts pressure on the gastro-splenic attachment which may be already weakened by the peritonitis and the attachment ruptures.

Splenoptosis is the displacement of part or the whole of the spleen from its normal location and can be primary or secondary in aetiology. Primary splenoptosis is often found in newborns and is therefore congenital, but it is not necessarily hereditary. Secondary splenoptosis is caused by the presence of concurrent disease and can occur any time in life. Splenomegaly is the enlargement of the spleen and loss of its sharp edges, which become rounded. Ectopic spleen refers to disjointed splenic tissue found in the parenchyma of other organs where it is not normally found, such as the liver or the pancreas. Ectopic spleen is also referred to as wandering spleen. Splenopexy is the surgical fixation of the spleen. Endoscopy is a surgical procedure that involves visual examination of internal body structures, in which an instrument is inserted in a body cavity for exploratory or corrective purposes. Finally, laparoscopy is a surgical procedure that involves examination of the peritoneal cavity in order to explore or correct ailments (Karvountzis, 2020).

Such a case is shown in Fig. 5.1. These images were taken during endoscopic reduction of a left displaced abomasum, in a cow with concurrent traumatic reticulitis. The images show exploratory findings during the abomasocentesis phase. As the volume of the displaced abomasum was being reduced, it revealed the true extent of the splenoptosis. It was found that the dorsal third of the affected spleen was partially displaced and attached through fibrinous adhesions to the displaced abomasum. The dorsal edge of the spleen was found to be pointing abaxially and ventrally, almost 180 degrees from its normal position, i.e. being placed dorsally and axially.

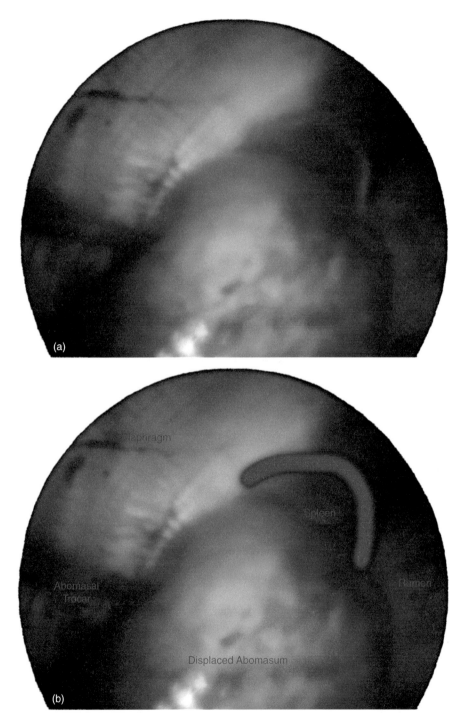

Fig. 5.1. Exploratory laparoscopy, where a case of splenoptosis is shown. Note the curved red line on (b), indicating the curvature of the partially displaced spleen. (Copyright Dr Sotirios Karvountzis.)

Left Displaced Abomasum (LDA)

LDA is common in cattle, occurring when the abomasum, the fourth compartment of the stomach, moves from its usual position on the right side of the abdomen to the left. It is a serious condition that can result in significant physiological and pathological changes in the animal and requires veterinary intervention.

Prompt diagnosis and surgical correction are typically necessary to restore the abomasum to its normal position and to address the associated pathological changes. Some of the key pathological changes associated with LDA in bovines are as follows:

Abomasal distention: The abomasum becomes distended with gas and fluid when it displaces to the left, a distention resulting from the accumulation of gases produced by bacterial fermentation of ingesta. This leads to the abomasum appearing to be inflated.

Vascular compromise: Displacement can result in stretching and displacement of blood vessels that supply the abomasum, leading to compromised blood flow, ischaemia – a lack of blood supply – and subsequent tissue damage.

Stomach wall oedema: The abomasal wall can undergo oedema, characterized by fluid accumulating in the tissues. Oedema occurs because of altered blood flow and the increased vascular permeability caused by the displacement.

Metabolic imbalance: Metabolic imbalances can result from the LDA disrupting the normal digestive process. The abomasum is responsible for acid secretion to aid in digestion and, when displaced, this acid production can be affected, resulting in a decrease in gastric acidity and disturbances in the overall digestive process.

Gastrointestinal stasis: Displacement can cause stasis or slowed movement of ingesta through the gastrointestinal tract, leading to further fermentation and gas accumulation which intensifies the distention.

Secondary inflammation: If the displacement and compromised blood supply are protracted, there can be secondary inflammation of the abomasal wall, leading to tissue damage, ulceration and even perforation of the abomasum.

Systemic effects: An LDA can generally impact on the animal's overall health as metabolic imbalances and inflammatory responses can lead to changes in electrolyte balance and acid-base disturbances, and alter blood parameters.

Right Displaced Abomasum (RDA)

RDA without volvulus is commonly seen in cattle, characterized by the abnormal displacement of the abomasum to the right side of the abdomen. This typically occurs in dairy cows during the early postpartum period. RDA without volvulus is less severe than RDA with volvulus but it can nonetheless lead to significant pathological changes.

Some of the key pathological changes associated with RDA without volvulus in bovines are as follows:

Abomasal distension: As a result of its abnormal location on the right side of the abdomen, the abomasum becomes distended primarily owing to an accumulation of gas, fluids and ingesta.

Impaired abomasal motility: The normal motility of the abomasum can be disrupted by its displacement. The muscular contractions needed for correct digestion and movement of ingesta through the abomasum are compromised, leading to impaired abomasal function.

Accumulation of fermentation gases: With impaired motility, there can be a failure by the abomasum to expel the fermentation gases that are generated during the normal digestion process. Gas then gathers within the abomasum, contributing to its distension.

Metabolic imbalances: RDA without volvulus can result in metabolic imbalances because of changes to digestion and nutrient absorption. The cow's ability to absorb nutrients adequately can be affected by the displacement and that means she can become deficient in essential substances such as electrolytes, proteins and vitamins or there can be an imbalance of these.

Abomasal ulcers: If the displacement of the abomasum is prolonged, there can be pressure necrosis and ischaemia in the affected region, leading to the development of abomasal ulcers, which compromise the overall health of the abomasum.

Systemic effects: Systemic effects such as metabolic acidosis, dehydration, reduced milk production, weight loss and general weakness can occur.

Abomasal Volvulus

Abomasal volvulus, commonly known as twisted stomach or displacement of the abomasum, routinely affects cattle. It is a severe condition that involves the abnormal twisting or displacement of the abomasum and can lead to significant pathological changes within the affected abomasum.

Some of the key pathological changes that occur during right abomasal volvulus are as follows:

Distension: The twisted abomasum becomes distended with gas, fluid and ingesta. The normal passage of ingesta is prevented by the rotation and obstruction of the abomasum, and that leads to the accumulation and distension of its contents. The abomasum can become greatly enlarged due to the accumulation of these gases and fluids.

Vascular compromise: The twisting of the abomasum can result in compression and twisting of the blood vessels that supply the abomasum, restricting blood flow to it. This leads to ischaemia and subsequent tissue damage. In severe cases the compromised blood flow can also contribute to the development of necrosis – tissue death.

Oedema and congestion: Oedema and congestion can result from the compromised blood flow and the accumulation of fluid within the abomasum – oedema due to increased vascular permeability and impaired fluid drainage resulting in the accumulation of fluid within the abomasal wall and lumen, and congestion from the pooling of blood within the abomasum due to impaired venous drainage.

Tissue necrosis: Tissue necrosis can be seen following prolonged ischaemia caused by the twisted abomasum because the lack of oxygen and nutrients leads to cell death, primarily affecting the inner layers of the abomasum. Necrotic tissue can be dark or black and have a foul odour.

Inflammation: The pathological changes in the abomasum prompt an inflammatory response. Inflammatory cells penetrate the affected tissues, releasing inflammatory mediators. This process further worsens tissue damage and contributes to the systemic effects associated with abomasal volvulus.

Peritonitis: The compromised abomasal wall can rupture in severe cases, causing the contents of the abomasum to spill into the abdominal cavity. This leads to peritonitis, the inflammation of the membrane lining the abdominal cavity, the peritoneum. Peritonitis is a life-threatening complication and can cause severe systemic illness.

Traumatic Reticulitis

Cattle are known to ingest foreign objects such as nails, wire or other sharp metallic objects, and when they do traumatic reticulitis, or hardware disease, can occur when the reticulum, one of the four compartments of the cow's stomach, is damaged. The sharp object can penetrate the reticulum, causing trauma to the surrounding tissues and leading to a variety of pathological changes. These include:

Inflammation: The initial response is inflammation – the damaged tissues release inflammatory mediators, and redness, swelling and increased blood flow in the affected area can follow.

Abscess formation: The inflammatory response in some cases can progress to abscesses forming. These localized pockets of pus develop as the body tries to isolate and contain the infection caused by the foreign object.

Fibrosis: Chronic or severe cases of traumatic reticulitis can result in fibrous tissue forming. Fibrosis is a reparative process in which the body attempts to replace damaged tissues with scar tissue. Normal tissue function can be lost and the normal architecture of the reticulum disrupted.

Adhesions: Adhesions, bands of fibrous tissue that can develop between different organs or between organs and the body wall, may form as part of the healing process. These can impede the movement and function of the affected organs and cause further complications.

Peritonitis: The foreign object in some cases can penetrate the reticulum, causing a perforation and a leakage of the stomach contents into the abdominal cavity. This leads to peritonitis, a life-threatening condition that requires immediate veterinary intervention. The severity and extent of internal pathology during traumatic reticulitis will be influenced by the size, shape and location of the ingested foreign object, and also the duration of the disease.

Abomasal Ulcer

Abomasal ulcers, or gastric ulcers, commonly affect bovines. These ulcers occur from the erosion or ulceration of the abomasum lining and can lead to internal pathologies within the affected abomasum. Some of the key pathological changes associated with abomasal ulcers in bovines are as follows:

Erosion and ulceration: The principal pathology of abomasal ulcers involves the development of erosions or ulcers on the inner lining of the abomasum.

Inflammation: An inflammatory response in the surrounding tissues is triggered by the presence of ulcers. Inflammatory cells, such as neutrophils and lymphocytes, infiltrate the damaged area, leading to inflammation. Chronic inflammation can contribute to further tissue damage and also impair healing.

Haemorrhages: There can be bleeding due to the erosion of blood vessels within the abomasal wall, which introduces blood in the abomasum and can result in anaemia in severe cases.

Perforation: Abomasal ulcers can develop to the point of perforation, where they extend through the entire thickness of the abomasal wall. Abomasal contents escape into the abdominal cavity, leading to peritonitis.

Fibrosis and scarring: Ongoing ulceration and healing attempts can result in the development of fibrosis and scarring within the abomasum. This fibrous tissue can replace the normal glandular tissue, impairing function in the affected region and impacting on digestion and nutrient absorption.

Secondary complications: Abomasal ulcers can have systemic effects on the animal's health, with chronic ulceration leading to weight loss, poor appetite, reduced milk production in dairy cows and productivity reduced overall. Animals that are affected can also have greater susceptibility to secondary infections and metabolic imbalances.

The development and severity of internal pathology during abomasal ulcers can vary depending on several factors, including the underlying causes, duration and overall health status of the affected animal.

Intestinal Ulcer

Intestinal ulcers, also known as bovine ulcerative jejuno-ileitis, or simply cattle ulcers, are a significant health issue in livestock production. They typically occur in the small intestine, particularly the jejunum and ileum. The exact cause is not fully understood but contributory factors are believed to include diet, stress, bacterial infections and changes in the gut microbiome. The internal pathology of intestinal ulcers involves many characteristic features:

Lesion formation: Ulcerative lesions of different sizes and depth appear as erosions or ulcers on the inner lining, the mucosa, of the small intestine. These range from superficial erosions to deep and penetrating ulcers.

Inflammation: The presence of ulcers causes an inflammatory response in the surrounding tissues with inflammatory cells, such as neutrophils and lymphocytes, infiltrating the affected area and causing more damage to the intestinal wall.

Haemorrhages: Ulceration can cause bleeding from the damaged blood vessels in the intestine with haemorrhage seen as blood in the faeces or as blood clots adhered to the ulcer site.

Oedema: Oedema can occur in the submucosal layer beneath the ulcer as a result of increased vascular permeability and leakage of fluid into the affected tissues.

Fibrosis and scarring: When ulcers are chronic or recurrent, scar tissue forms during the healing process. In response to the ulceration, fibrosis occurs as a reparative response, thickening and remodelling the affected area.

Complications: The consequences of intestinal ulcers can be severe for bovines. They can cause malabsorption and malnutrition by disrupting the normal absorptive function of the intestine and their presence can also increase the animal's susceptibility to secondary infections or further complications, such as intestinal obstruction or perforation.

The diagnosis of intestinal ulcers typically involves a combination of clinical signs, decreased appetite, weight loss and diarrhoea, for example, and, occasionally, blood in the faeces. Endoscopic examination of the intestinal tract can confirm the presence of ulcers and assess their severity.

Kidney Issues

Renal disease, often caused by ingesting toxic plants or chemicals, is the most common internal pathology of kidney issues in bovines. There are some key aspects of the internal pathology associated with kidney issues in bovines:

Nephritis: This refers to the inflammation of the kidney tissue, occurring as a result of bacterial or viral infections, toxins or immune-mediated diseases. Leptospirosis, caused by the *Leptospira* species, is a common infectious cause of nephritis in bovines.

Tubulointerstitial nephritis: This condition, involving the inflammation of the tubules and interstitial tissue of the kidneys, can be caused by factors including toxins, medications, infections or autoimmune diseases. Mycotoxins produced by fungi present in contaminated feed can contribute to tubulointerstitial nephritis in bovines.

Pyelonephritis: Pyelonephritis, a bacterial infection affecting both the renal pelvis and the kidney tissue, commonly arises from ascending urinary tract infections, where bacteria enter the kidneys from the bladder or ureters. Urinary tract obstruction, bladder dysfunction or prolonged use of indwelling catheters can lead to pyelonephritis.

Renal cysts: These fluid-filled sacs form within the kidney tissue and can be congenital or acquired. Acquired cysts can result from polycystic kidney disease, chronic kidney disease or other genetic abnormalities.

Renal calculi: Also known as kidney stones, these are solid deposits that form within the kidneys where they can obstruct the urinary tract, causing pain, inflammation and damage to the renal tissue. Causes include dietary imbalances, mineral deficiencies or excesses and inadequate water intake.

Renal tumours: Although renal tumours are relatively rare in bovines, they do occur and the most common among these is renal cell carcinoma that arises from the cells lining the renal tubules. Structural damage to the kidney can result, as can impaired renal function.

Metritis

Metritis is an inflammation of the uterus and commonly occurs after calving. It can have significant economic consequences in dairy cattle due to decreased fertility, increased culling rates and reduced milk production. Several changes in the uterus contribute to the clinical signs and complications associated with this condition.

Uterine inflammation: An inflammatory response within the uterus is characteristic of metritis. The uterus lining, the endometrium, thickens and becomes congested with inflammatory cells, such as neutrophils and macrophages. A typical cause of the inflammation is bacterial contamination of the uterus following calving.

Uterine wall thickening: When metritis becomes chronic, the uterine wall can thicken, while prolonged inflammation and repeated infections can lead to fibrosis within the uterine tissue. Fertility is further impaired by the thickened uterine wall reducing the elasticity and normal function of the uterus. Metritis can have different degrees of severity, with the specific internal pathology varying between individual cows.

Umbilical Hernia Explorations

The purpose of endoscopically examining umbilical hernias is to establish the extent of the formation of adhesions between the lining of the hernial sac and the mesenterium or the intestine. Such examination is of significant prognostic value in the treatment of the hernia.

To carry out the operation, the animal must be in dorsal recumbency and all the handling and tranquilizing considerations of the two-step technique should be taken into account. When exploring the abdomen or thorax, we can use the 5-mm or 10-mm magnetic trocars for the 5-mm and 10-mm rigid endoscopes, respectively. There are fixed and portable light sources that can be used in the field, with a number of considerations accompanying them.

Light Sources Used for the Procedure

The fixed light sources produce high-quality internal images with up to 2000 lumens of luminosity. The disadvantage of these is that they need to be plugged into mains electricity, which is not always available or safe in cattle accommodation. They are usually quite heavy to carry and to store securely while they are being used. Finally, a good-quality light source is expensive to purchase, another consideration when carrying it in the field.

A portable light source is less expensive than a fixed one, easier to carry and most are powered by rechargeable batteries. Having a number of spare rechargeable batteries to hand is a sensible back-up plan, minimizing interruptions during patient examinations. The disadvantage of a portable light source is its luminosity. At the time of writing, a reasonable portable light source produces 200 lumens of light.

Most exploratory images during a laparoscopy or thoracoscopy in this publication unless stated otherwise were collected with a portable light source, powered by rechargeable batteries, producing approximately 200 lumens of light.

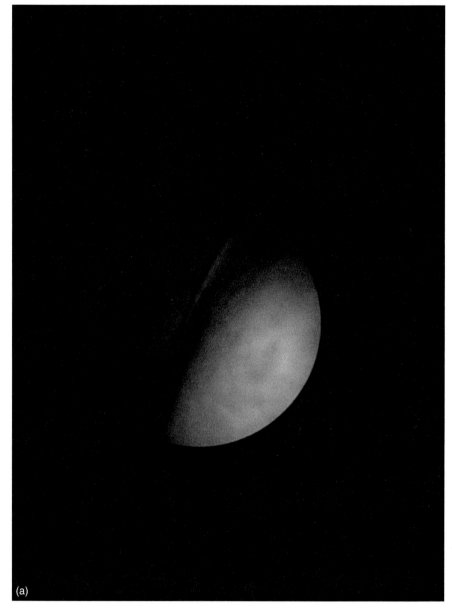

(a)

Visualization of the Exploratory Laparoscopy

Fig. 5.2 and is an image taken during exploratory laparoscopy showing a healthy spleen. It is paramount for the endoscopic surgeon to train their eye on the visual characteristics of a healthy organ, as well as its pathology, during endoscopic examinations.

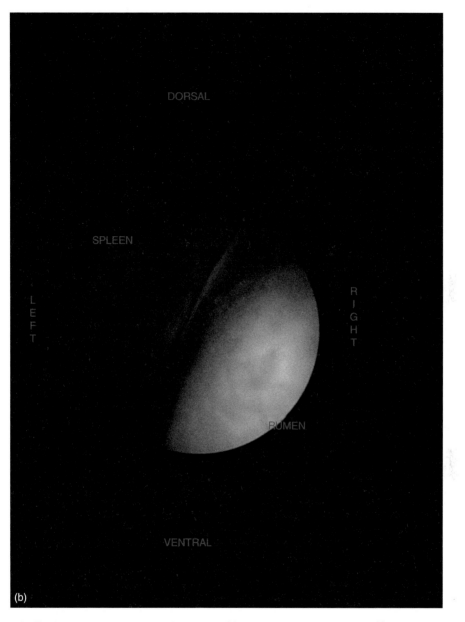

Fig. 5.2. Exploratory laparoscopy, where a healthy spleen is demonstrated. (Copyright Dr Sotirios Karvountzis.)

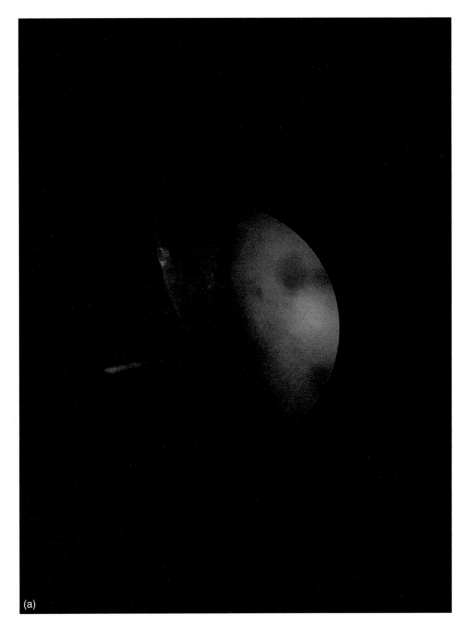

(a)

Fig. 5.3 is an image taken during a left displacement of abomasum correction with the standing technique and shows a healthy spleen, ecchymosis on the rumen serosa and the distal end of the abomasal cannula. It is a fundamental advantage of endoscopic applications that in most cases we can fully visualize the insertion of sharp instruments internally and, most of the time, prevent unintended visceral trauma.

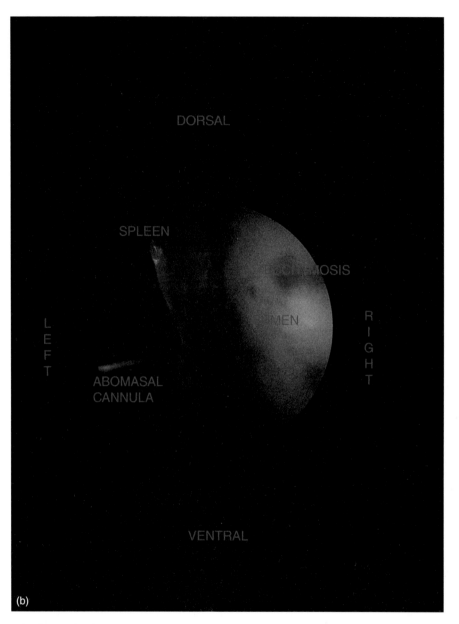

Fig. 5.3. Corrective laparoscopy, where the advance of the abomasal trocar is demonstrated. (Copyright Dr Sotirios Karvountzis.)

Fig. 5.4. Position of the surgeon and handling of the magnetic trocar, when setting up the optical portal. (Copyright Dr Sotirios Karvountzis.)

The most important part of any endoscopic procedure is preparation, starting with animal restraint. Most endoscopic procedures can conclude swiftly, provided the patient is controlled properly. The best location to restrain an animal that is going to be operated on is a corner of a handling pen. This position limits the patient's lateral and sagittal movements, contributing towards a smooth operation.

When setting up the optical portal, for example, to carry out a corrective laparoscopy of a left displacement of the abomasum, the surgeon places the 5- or 10-mm trocar through the skin incision, where the portal will be set. The insertion of the instrument through the remaining abdominal wall then takes place, as shown in Fig. 5.4.

Fig 5.5. Once the optical trocar has been established, the surgeon checks the flow of gases and also their smell. (Copyright Dr Sotirios Karvountzis.)

One hand stabilizes against the ribs and the other hand holds the magnetic trocar, which is already through the skin incision. With the handle of the awl in the surgeon's palm, we press in a single sharp motion aiming axially and dorso-cranially.

During this step, the valve of the gas vent remains in the 'off' position, ensuring that no introduction of air has taken place so far. Once the trocar is in place, we switch the valve to the 'on' position, allowing a flow of gases, as shown in Fig. 5.5.

Fig. 5.6. Unscrewing the seal of the magnetic trocar to enhance passive insufflation of the abdomen. (Copyright Dr Sotirios Karvountzis.)

The flow of gases can be checked by sound, but also by spraying some liquid on the inlet of the gas vent valve. The liquid will spray out, in the case of out-flowing gas, or will be sucked into the cannula, in the case of insufflation. This check is paramount, as if the trocar is in the peritoneum, which operates under negative air pressure, the tendency is for gases or liquids to flow into the body. In the case of unintended visceral paracentesis, for example the abomasum or the rumen, these organs operate under positive air pressure and the tendency would be for those gases to flow towards the surgeon.

Once the surgeon is certain that the optical portal has been set up correctly, the seal of the cannula can be unscrewed in order to allow a copious amount of air to be introduced into the peritoneum, as shown in Fig. 5.6. This constitutes passive insufflation whereas, when needed, active insufflation with a cordless pump can also be used.

The next step would be to carry out a visual inspection of the area. This endoscopic exploration is of the utmost importance as it will aid our diagnosis, and also our prognosis of the case. The surgeon is encouraged to gather as much evidence as possible by viewing many aspects of the area. For example, in exploratory laparoscopy of the left flank, we start by aiming the endoscope cranially, then rotate to caudally as shown in Fig. 5.7.

Fig. 5.7. Carrying out exploratory endoscopy of the left flank. The surgeon starts by inspecting the cranial abdomen first (a) and then rotates caudally for the retro-peritoneal organs (b). (Copyright Dr Sotirios Karvountzis.)

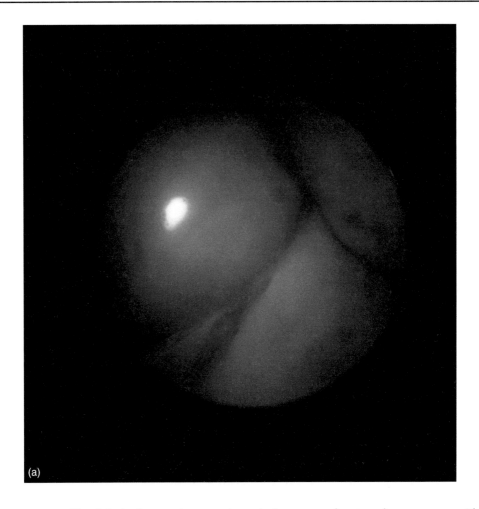

(a)

Fig. 5.8 depicts an image taken during an exploratory laparoscopy with the two-step technique, through an optical portal located in the ventral abdomen. It shows the pylorus and small intestine of a healthy patient. The advantage of the two-step exploratory technique is that it does not leave any organ unchecked, allowing the surgeon to establish with confidence the possible cause of the ailment.

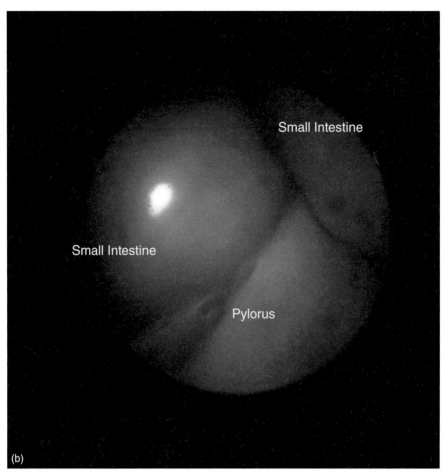

Fig. 5.8. Exploratory laparoscopy, visualizing a healthy pylorus and small intestine. (Copyright Dr Sotirios Karvountzis.)

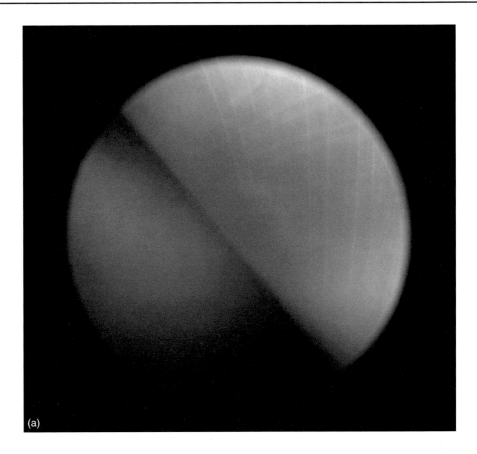

(a)

Fig. 5.9 shows an image taken during a corrective laparoscopy with the standing technique, through an optical portal located in the left paralumbar fossa. It shows the greater curvature of the left displaced abomasum, being in contact with the rumen. While carrying out endoscopic operations, exploratory or corrective, it is important to assess the colours of the visualized organs. In this case, the abomasal mucosa appears slightly white or yellowish and the ruminal mucosa mostly white. These are normal colourings of these organs, indicating that the ailment is relatively fresh. This in turn indicates a likely positive prognosis for the patient.

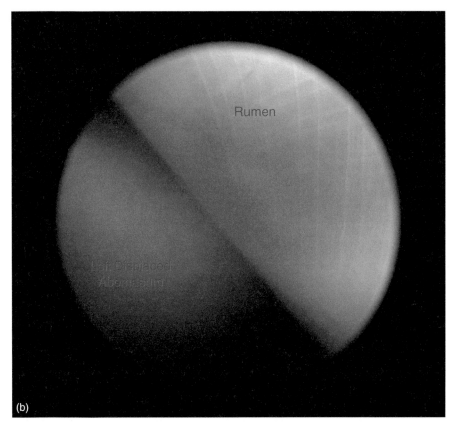

Fig. 5.9. Corrective laparoscopy, showing the left displaced abomasum and the rumen. (Copyright Dr Sotirios Karvountzis.)

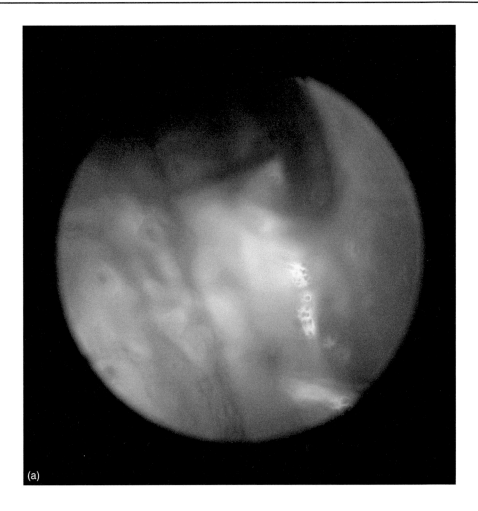

(a)

Fig. 5.10a and b are images taken during an exploratory laparoscopy with the two-step technique, through an optical portal located in the ventral abdomen. They show the body of the normal abomasum, lying next to the omasum. Note the colouring of both organs, slightly pink for the abomasum and white for the omasum, indicating their healthy state.

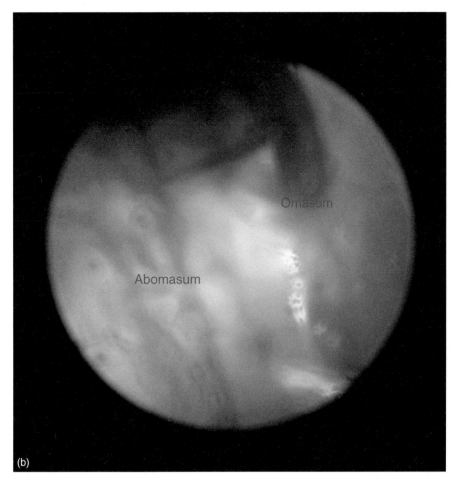

Fig. 5.10. Exploratory laparoscopy, depicting a healthy abomasum adjacent to a healthy omasum. (Copyright Dr Sotirios Karvountzis.)

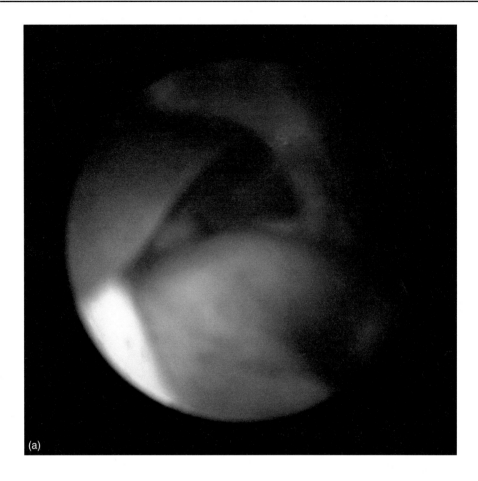

(a)

Fig. 5.11a and b are images taken during an exploratory laparoscopy with the two-step technique, through an optical portal located in the ventral abdomen. It demonstrates the healthy liver of a cow lying next to the abomasum. The colour and sharp edges of the liver are notable, as the healthy organ is pale red and has sharp edges. In case of circulatory stasis, due to traumatic reticulitis, for example, the liver becomes congested and swollen, with its edges becoming rounded.

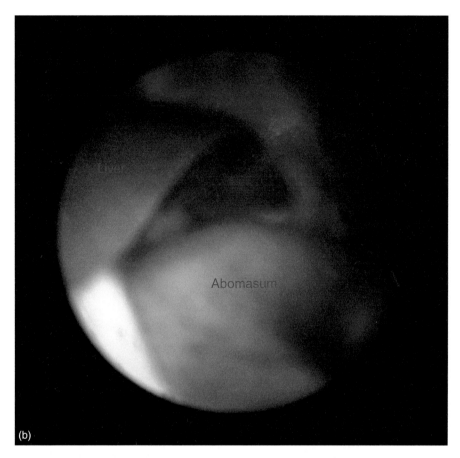

Fig. 5.11. Exploratory laparoscopy, showing a healthy liver situated next to a healthy abomasum. (Copyright Dr Sotirios Karvountzis.)

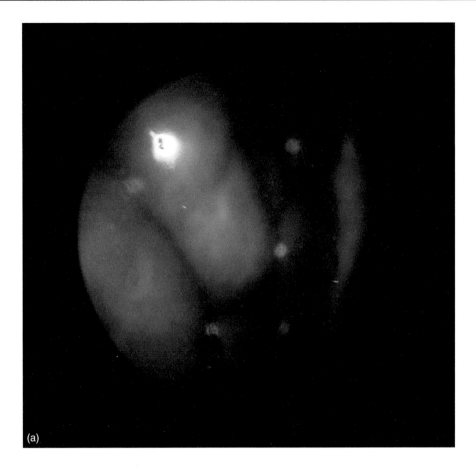

(a)

Fig. 5.12a and b are images taken during an exploratory laparoscopy with the two-step technique, through an optical portal located in the ventral abdomen. They indicate the inflamed small intestine, resulting from extensive peritonitis due to traumatic reticulitis. What is striking about these images is the colour of the small intestine, part of jejunum in this case, where the intestinal mucosa has altered to dark red. The colour of the normal small intestine may vary from pink to pale red depending on a variety of factors, but only in very rare cases would it reach such a red shade.

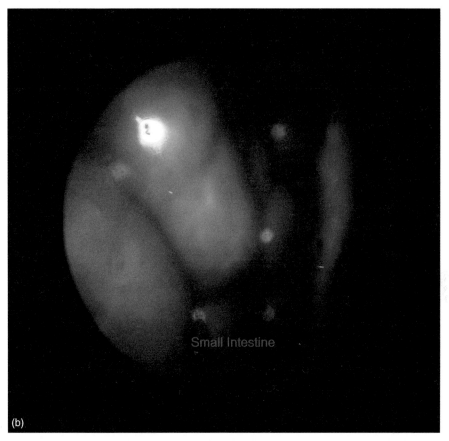

Fig. 5.12. Exploratory laparoscopy, demonstrating the inflamed small intestine, resulting from extensive peritonitis. (Copyright Dr Sotirios Karvountzis.)

(a)

Fig. 5.13a and b are images taken during an exploratory laparoscopy with the two-step technique, through an optical portal located in the ventral abdomen. They show the healthy liver of a patient with left displacement of the abomasum. Note the colour of the organ – light purple – indicating a healthy organ.

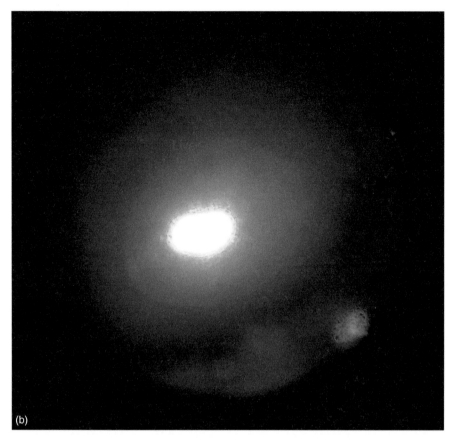

Fig. 5.13. Exploratory laparoscopy, representing a healthy liver. (Copyright Dr Sotirios Karvountzis.)

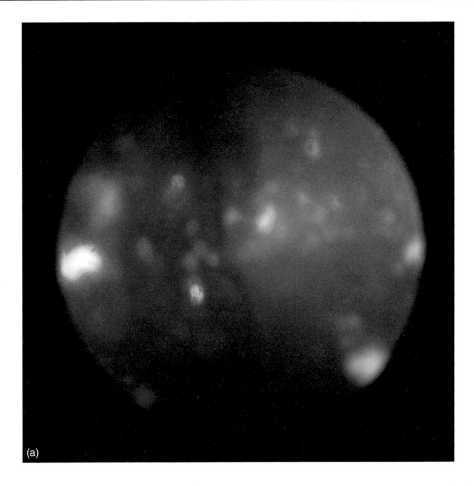

(a)

Fig. 5.14 shows images taken during an exploratory laparoscopy with the two-step technique through an optical portal that took place in the ventral abdomen. They demonstrate the typical white deposits of fibrin that are the result of extensive peritonitis. This image was taken over the liver. The likely cause of peritonitis in this case was traumatic reticulitis, due to a foreign body.

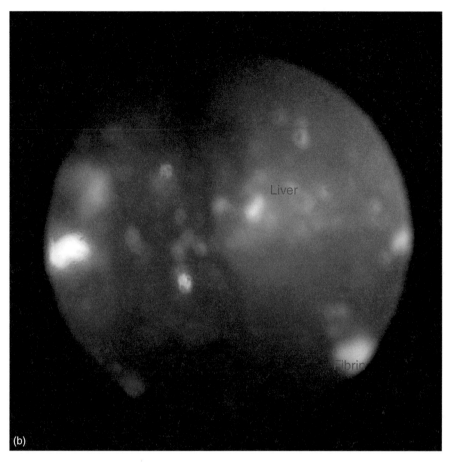

Fig. 5.14. Exploratory laparoscopy, showing the typical white fibrin deposits during peritonitis. (Copyright Dr Sotirios Karvountzis.)

6 Corrective Laparoscopy – Left Displacement of Abomasum

Abstract

This chapter explores the uses of corrective laparoscopy in cattle, focusing on the left displacement of abomasum. Common ailments, such as abomasal displacements, can be addressed without delay, rapidly and effectively, with a number of minimally intrusive techniques. Corrective laparoscopy brings a myriad of advantages to the bovine patient and a wealth of gratification to the endoscopic surgeon.

Laparoscopic corrections of abdominal ailments are commonplace in human surgery and are even carried out by robots or remotely. They are gradually becoming popular with cattle vets all over the world, because the surgical kit is becoming more affordable and the procedure can be carried out on the farm during a routine appointment or as part of an emergency call-out. The technique offers a less intrusive and more effective approach for correcting the left displacement of abomasum, compared with laparotomic techniques.

Standing Technique

Use a halter to position the patient in a corner of the pen or examination box. It should be tied to a secure attachment, such as the lower pins of the hanging gate. The assistant stands by the head, holding the animal's nose and halter when required.

Place shackles on the rear legs to avoid the surgeon being kicked during the operation. Shackles are positioned on the metatarsals, in order to stop the patient kicking. Ideally, shackles should have quick-release buckles, such as a belt, and not be made of Velcro or have a screw-lock mechanism, as these can release very easily, or not at all, during the operation.

Consider how you will immobilize the patient's tail. This is important because a flicking tail during the operation can not only introduce impurities into the operating field but can severely disrupt the surgeon during the procedure. There are two options, either attaching a string from the tail to one of the rear legs, or epidural anaesthesia. The advantage of the tail string is that it is simple and there are no considerations regarding meat or milk withdrawal periods. The disadvantage of the tail string is that it sometimes fails during the operation.

Tail epidural requires the injection of local anaesthetic that contains procaine hydrochloride and epinephrine into the spinal canal. A total of 2 ml is required and the duration of the effectiveness of the anaesthesia is usually between 1 and 1.25 hours. The advantage of this method is its effectiveness in immobilizing the tail. The disadvantage lies in the fact that at the time of writing no commercial injectable anaesthetic preparations are licensed in the UK for bovine epidural anaesthesia, therefore standard milk and meat withdrawal periods may need to be applied. The anaesthetic preparation is injected in the intervertebral space between cervical 3 and cervical 4 vertebrae or cervical 4 and cervical 5 vertebrae. My preferred approach for an epidural is the use of an 18-gauge 1.5-inch-long needle that is inserted at a 45-degree angle into the spinal column, without any skin preparation prior to the procedure.

Patients that will undergo endoscopy while standing may have to be sedated. I prefer xylazine hydrochloride as a tranquilizer administered intravenously, at a dose rate of 2–20 mg per animal. The lower dose rate is reserved for placid animals, whereas the higher one is for fractious patients. Depending on the patient's demeanour, the anaesthetic effect may last for up to 30 minutes.

The preparation of the portals involves depilation, local anaesthesia and disinfection. A square area of hair removal of 5–7.5 cm in size, surrounding the skin incision, is adequate. Depilation can be carried out by rechargeable clippers or by a shaving blade. The size of the clipper blades used is between size 10 and size 20.

For the local anaesthesia of the portals, between 5 and 10 ml per site of procaine hydrochloride and epinephrine can be used. The anaesthetic is injected subcutaneously as a 2–5 cm-long line block at the edge of the clipped area, intersecting with the route of the sensory nerves in the region.

Abomasocentesis

It is recommended that introducing the 13-mm magnetic trocar at the working portal site should be done while under full endoscopic observation. The surgeon is advised to guide the 13-mm magnetic trocar while observing its passage internally with the endoscope. This minimizes significantly unexpected visceral trauma and presents a significant advantage of this technique over other bovine exploratory or corrective surgical techniques, as shown in Fig. 6.1.

Fig. 6.1. Introducing the 13-mm magnetic trocar at the working portal site, best done under observation with the endoscope. (Copyright Dr Sotirios Karvountzis.)

When setting up the working portal, when we aim for abomasopexy near the navel, we select an area for setting up this portal that lies halfway down the left flank. We use the 13-mm magnetic trocar, insert it through the skin incision and aim dorso-axially or caudo-axially, depending on the rumen fill and the level of abdominal insufflation. One of the objectives here is to avoid unintended entry into the rumen, as shown in Fig. 6.2.

Fig. 6.2. Setting up one of the working portals for abomasopexy near the navel. (Copyright Dr Sotirios Karvountzis.)

Fig. 6.3. The start of the abomasocentesis: insertion of the abomasal trocar through the working portal, while under observation with the endoscope. (Copyright Dr Sotirios Karvountzis.)

Abomasocentesis – the puncturing of the abomasum – starts with the insertion of the abomasal trocar through the working portal. The working portal cannula is already in place and through it we insert the abomasal trocar. This step is done under observation and provides one of the major advantages of endoscopic correction, that is the full observation of sharp instruments while they are being handled in the abdomen. This in turn reduces the risk of inadvertent visceral trauma.

The surgeon handles the endoscope and light source, which is located through the optical portal, with the right hand, and the abomasal trocar, which is located through the working portal, with the left hand, as shown in Fig. 6.3. Being ambidextrous during this step is not a prerequisite, but it does help.

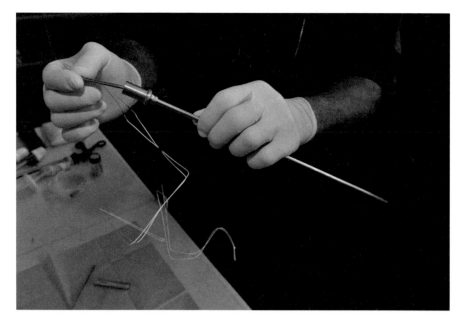

Fig. 6.4. Inserting the metal T-bar of the abomasal toggle into the abomasal trocar. (Copyright Dr Sotirios Karvountzis.)

Following the abomasocentesis, the surgeon needs to prepare for the insertion of the T-bar toggle into the abomasum, while the abomasal cannula is *in situ*. Place the metal bar of the T-bar toggle into the external opening of the abomasal cannula, as shown in Fig. 6.4.

Push the metal bar all the way into the abomasal cannula with your thumb and, with the aid of the toggle introducer, push it as far as the introducer can go. This is shown in Fig. 6.5.

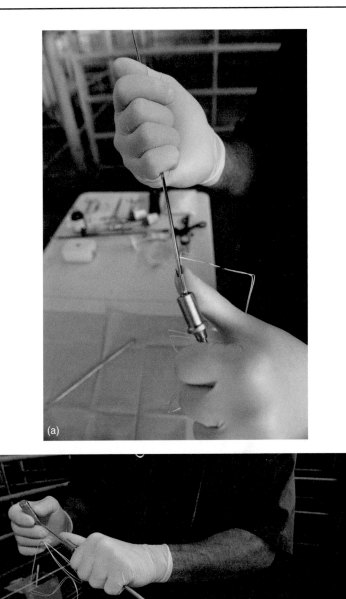

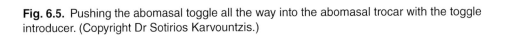

Fig. 6.5. Pushing the abomasal toggle all the way into the abomasal trocar with the toggle introducer. (Copyright Dr Sotirios Karvountzis.)

Fig. 6.6. The transfixer is the longest instrument included in an endoscopic kit. Suggested clearance above the surgeon is indicated here, so that no obstructions exist above the animal. (Copyright Dr Sotirios Karvountzis.)

Before the introducer is retracted from the abomasal cannula, ensure that you re-insert the introducer three times, until the introducer's metal handle touches the outer rim of the abomasal cannula and the distinct 'clicking' sound is heard. This step is important, as sometimes the metal T-bar of the toggle comes into contact with serum or blood in the lacuna of the abomasal introducer and remains in it when the abomasal cannula is eventually retracted from the abomasum.

Before inserting the transfixer through the working portal, the surgeon needs to bear in mind a few safety aspects connected to its handling. Firstly, this particular instrument is quite lengthy, therefore considerable clearance is needed above the animal to ensure that the managing of the instrument will not interfere with the existing animal handling facilities. The suggested clearance required is shown in Fig. 6.6.

Fig. 6.7. There is the risk of injury to the surgeon if the transfixer is carried with the handle pointing away, even if the transfixer needle is retracted. (Copyright Dr Sotirios Karvountzis.)

Secondly, when carrying the assembled transfixer on the farm, care is needed to avoid injury to the surgeon or other staff. Because the instrument is quite long, the transfixer's needle must be retracted in between uses, in other words with its sharp tip placed inside the transfixer's lacuna. If the surgeon is carrying the transfixer with the handle end of the transfixer pointing away, there is the risk of injury to the surgeon when the handle end accidentally comes into contact with an unexpected obstacle. Avoid carrying the assembled transfixer, as shown in Fig. 6.7.

Fig. 6.8. Two transfixers for abomasopexy. The transfixer shown in (a) is used for pexy near the xiphoid process and the instrument in (b) is used for pexy near the navel. The latter instrument has a larger curvature to its distal end. (Copyright Dr Sotirios Karvountzis.)

Depending on whether the abomasopexy will take place near the xiphoid process or near the navel, a different transfixer needs to be used. The main difference is in the curvature of the distal end of this instrument. There are two transfixers, one designed for abomasopexy caudally to the xiphoid and one for abomasopexy cranially to the navel, as shown in Fig. 6.8. The transfixer that is used for abomasopexy near the navel has a larger curvature because it has to cross the abdomen underneath the rumen; the transfixer's curvature needs to match the curvature of the ventral sac of the rumen as much as possible.

When inserting the transfixer through the working portal, care needs to be taken for adequate clearance above where the patient is restrained. An ideal set-up is shown in Fig. 6.9.

Fig. 6.9. Clearance for two types of transfixers used for abomasopexy. The instrument is quite lengthy and good clearance is required, particularly in the case of the transfixer used for abomasopexy near the xiphoid process, as shown in (a). (Copyright Dr Sotirios Karvountzis.)

Before exteriorizing the needle of the transfixer, particularly when the abo-masopexy takes place near the xiphoid process, the instrument needs to be facing the correct way. As has been elaborated earlier, the handle of the trans-fixer consists of two parts of unequal length. The longer part of the transfixer handle indicates the direction that the curved tip of the transfixer (which lies inside the cow and is not visible to the surgeon) points toward. When trans-versing through the abdomen, ensure that the longer part of the transfixer's handle, which is visible and above the working portal, is pointing axially. This is because transversing is easier when the curvature of the transfixer matches the curvature of the abdominal wall.

When the transfixer is *in situ* and we are ready to exteriorize the needle through the abdominal wall at the point of abomasopexy, we need to reverse this configuration. The transfixer's needle exteriorizes with more ease when the curvature of the distal part of the transfixer is at a right angle to the abdominal floor. In this case, we need to rotate the handle of the transfixer by 180 degrees so that the longer part of the handle will be pointing abaxially. The two different settings of the transfixer handle are shown in Fig. 6.10.

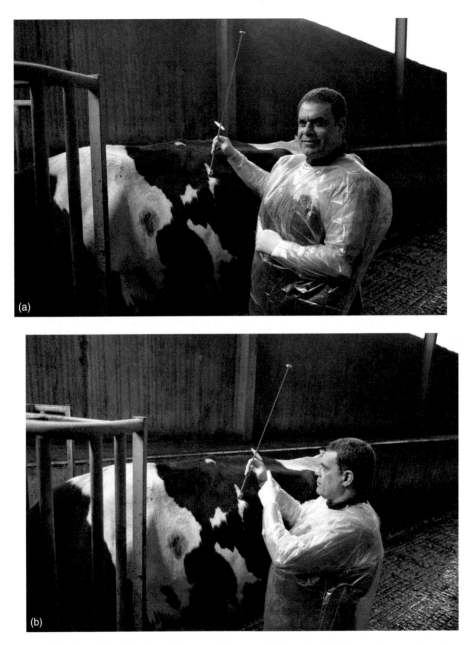

Fig. 6.10. The handle of the transfixer provides us with invaluable information on how the instrument lies inside the abdomen during standing correction of a left displacement of the abomasum. As the transfixer migrates to the abomasopexy point near the xiphoid process, the long part of the handle needs to point axially as shown in (a). When the transfixer is *in situ* and ready for its needle to be exteriorized, the instrument needs to rotate 180 degrees with the longer part of the handle pointing abaxially (b). (Copyright Dr Sotirios Karvountzis.)

Fig. 6.11. Configuration of the transfixer when abomasopexy takes place near the navel. Please note the handle of the transfixer has been deliberately connected in a sagittal manner in relation to the animal, as rotation of the instrument does not take place with this method. (Copyright Dr Sotirios Karvountzis.)

Such a configuration of the transfixer handle is not a concern when it comes to abomasopexy near the navel. The transfixer used has a larger curvature to match and to slide underneath the curvature of the rumen. The room for manoeuvre is much more limited in this case and once the instrument is *in situ* exteriorization of its needle can take place without rotating the handle. The configuration of a transfixer for abomasopexy near the navel is shown in Fig. 6.11, where the handle has been deliberately connected in a sagittal manner in relation to the patient.

When migrating the transfixer during abomasopexy near the xiphoid process, from the working portal that lies dorso-left to the abomasopexy point that lies ventro-right, the surgeon is advised to hold the handle of the transfixer with the right hand, lean into the standing patient with the left shoulder and finally use the left hand to touch the abdominal skin. The function of the left hand is paramount here as we need to ensure no viscera is trapped at the bulb of the distal end of the transfixer that lies internally and the needle of the transfixer when it is exteriorized. This is achieved by ensuring our left hand externally is in constant contact with the transfixer's bulb internally, until such time as it reaches the abomasopexy point and its needle is exteriorized. The surgeon's configuration is shown in Fig. 6.12.

Fig. 6.12. Posture of the surgeon while the transfixer migrates to the abomasopexy point near the xiphoid process. Note the placement of the right hand on the handle of the transfixer, the left shoulder against the patient and the left hand following the bulb of the distal end of the transfixer. (Copyright Dr Sotirios Karvountzis.)

Abomasopexy

Before we load the extra piece of nylon suture to the transfixer, we need to create a loop at one end, as a scaffold or an easy loop. The other end has a pair of forceps attached, to stop the free end of the nylon suture inadvertently being pulled in the abdomen. Once the nylon suture and the T-bar toggle sutures are connected, the nylon suture's free end will be used to pull the T-bar sutures through the fixation point and have them fixed permanently there. The loop is then collapsed and threaded through one of the holes in the transfixer needle while this needle is exteriorized through the transfixer cannula, as shown in Fig. 6.13.

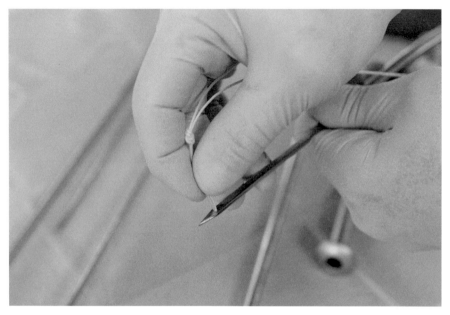

Fig. 6.13. In preparation for the abomasopexy, the extra loop of nylon suture is threaded through the eye of the transfixer needle. (Copyright Dr Sotirios Karvountzis.)

The collapsed loop is pulled through the transfixer needle hole until the knot of the loop blocks the needle hole. The collapsed loop on the other side is opened up and the remaining nylon sutures are threaded through the re-opened loop, as shown in Fig. 6.14.

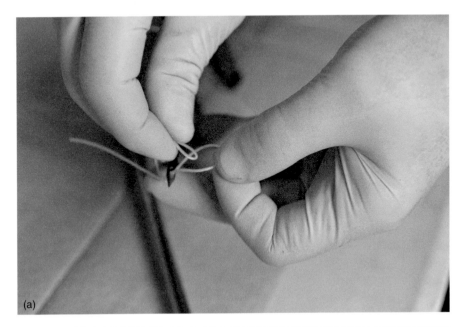

(a)

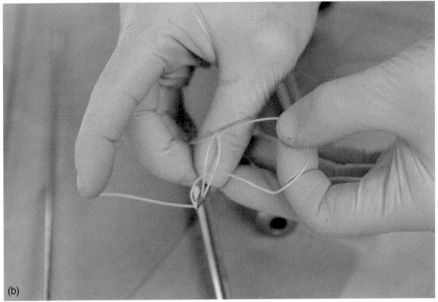

(b)

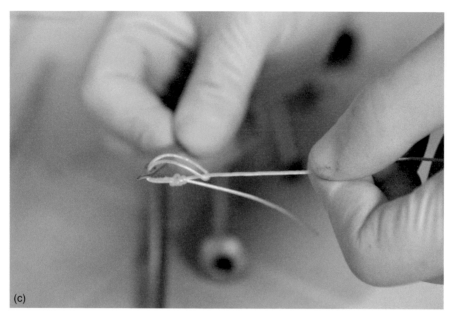

(c)

Fig. 6.14. Introducing and securing the extra piece of nylon suture at the needle of the transfixer. (Copyright Dr Sotirios Karvountzis.)

Once all the suture is pulled through the threaded loop, the knot of the extra piece of nylon suture is now on the side of the transfixer needle. The needle is then retracted through the transfixer cannula and this pulls the knotted nylon suture with it, as shown in Fig. 6.15.

Fig. 6.15. Retracting the needles of the transfixer with the extra piece of nylon suture attached to it, in order to transport it across the abdomen safely. (Copyright Dr Sotirios Karvountzis.)

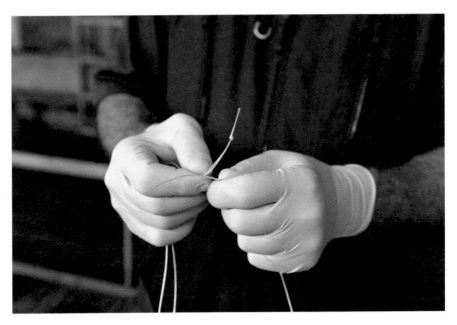

Fig. 6.16. Extending the nylon sutures of the T-bar toggle with the extra piece of nylon suture, which was transported from the abomasopexy point with the transfixer. (Copyright Dr Sotirios Karvountzis.)

The transfixer with the retracted needle and the attached nylon suture is then pulled through the animal and exteriorized through the working portal. Once the nylon sutures are recovered through the working portal, they need to be connected with the double nylon sutures of the T-bar toggle, as shown in Fig. 6.16.

Fig. 6.17. Securing the extended suture with multiple simple knots, while ensuring that each strand of that knot is secure. (Copyright Dr Sotirios Karvountzis.)

This is done by multiple simple knots, while ensuring that each of the four strands of the resulting knot are secure and are not slipping, as shown in Fig. 6.17. Once we have connected the nylon suture with the nylon strands of the T-bar toggle, our extended suture is complete. One end of this thread contains the metal T-bar and is found in the abomasum; the middle part of this thread lies outside the working portal with the interconnecting knot; and the final part of this thread is left exteriorized at the fixation point and held securely with a pair of artery forceps.

Upon completion of the suture extension, we conclude the fixation of the abomasum by using a pair of artery forceps to pull on the nylon suture that is secured. The extended suture is slowly exteriorized until the connecting knot between the nylon suture and the T-bar toggle sutures comes through. We need to pay particular attention at this point as, depending on the size of the knot, a gentle tug is usually enough for the remaining suture to come through the abdominal wall.

The loose T-bar toggle sutures need to be placed in a way that does not harm the patient's skin post-operatively. If the suture ends were to be tied in a knot as a means of securing them, leaving this knot in direct contact with the skin may lead to a fistula at the site of the abomasopexy. This is a result of the continuous pressure applied on the knotted sutures externally from the constant peristalsis of the contracting abomasum while it fulfils its physiological purpose of digestion. The bare sutures would cut through the abdominal wall.

A simple solution is to thread the loose T-bar toggle sutures through a rolled bandage and then tie a knot, with the bandage placed between the skin and the knot, as shown in Fig. 6.18. This will protect the skin from the constant 'tugging' of the abomasum. The disadvantage of using a bandage is its absorbent properties, as it will collect impurities post-operatively, particularly as the cow is resting in the housing accommodation. An alternative to the bandage is a purpose-built reusable protector.

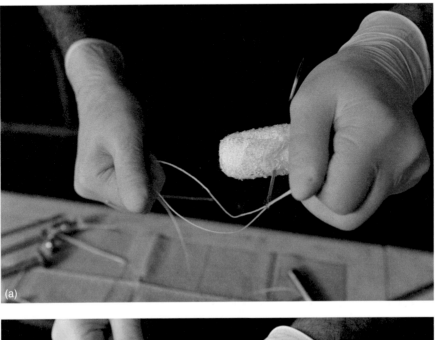

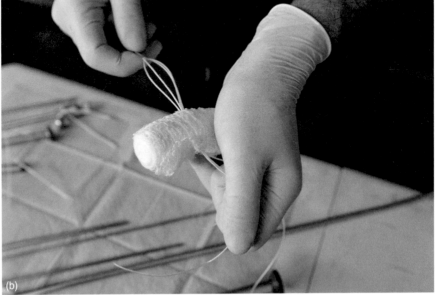

Fig. 6.18. Introducing a rolled bandage at the site of abomasopexy, as a means of protecting the skin from suture abrasion damage. (Copyright Dr Sotirios Karvountzis.)

Two-step Technique

Once the animal is ready for dorsal recumbency, the following should be considered in addition to the above preparations.

A second set of shackles that are positioned on the metacarpals should be used. A rope that runs sagitally along the body connects the two sets of shackles. This arrangement can be quite useful, not only in casting the animal, but more importantly in restraining the patient while cast. Two, but preferably three, assistants should be considered, to assist with the safe rolling of the animal.

If the patient is going to remain recumbent for more than 20 minutes, it is advisable to intubate with an oesophageal tube to prevent regurgitation of ruminal contents and aspiration pneumonia. Alternatively, tracheal intubation may be considered.

Unless there is a purpose-built crush for carrying out the rolling and restraint of the animal, a 12-m length of rope can be used for casting the patient. The casting may be carried out with the Reuff or the Burley methods. The important aspect in both of these casting methods is to remove any slack in the rope before traction is applied.

Once the animal is in dorsal recumbency, it needs to be restrained while its right ventral side is slightly higher than its left ventral side. This can be achieved by propping the animal up against a wall or a cubicle rail separator, or by hooking the front-end loader arm of a tractor on the sagittal rope that connects the two shackles and lifting the cow slightly off the ground, or by placing the animal in the front-end bucket loader of a tractor.

The patient may need to undergo heavy sedation or general anaesthesia for the two-step technique to be carried out. Heavy sedation can be reached by administering 30–50 mg per animal of xylazine hydrochloride intravenously. If a general anaesthetic regime is required, then the combination of xylazine hydrochloride and ketamine injected intravenously is a valid alternative. The dose rate is xylazine 0.14–0.22 mg/kg, followed by ketamine 2–5 mg/kg. The onset of sedation is usually within one minute and the effectiveness of the anaesthetic may last for up to 30 minutes.

The patient should be placed in the farm's hospital bay post-operatively for 3 days, while joining the herd during normal milk routine. Feed intakes and milk yields should be monitored, although the latter should not be expected to make a full recovery before 10 days post-op. Oral rehydration of the patient with a large volume of fluids (20–40 l) is advisable immediately after the operation. Choose an analgesic regime with the impact on the milk withdrawal period in mind. No antibiosis is required, unless the patient is presenting with bacterial concurrent disease. In such cases, treat accordingly using a 3-day broad-spectrum antibiotic course. Due to the particularly small size of the portals and the speed at which they normally heal, no considerations are required here. The application of antibiotic spray is complementary, but not proven to be beneficial. Finally, remove the gauze roll placed at the abomasopexy point within 4–6 weeks of the operation.

Provided the patient was free of concurrent disease pre-operatively, most post-operative complications for endoscopic techniques are predictable and relatively easy to prevent. Poor restraint, particularly when the patient is confined anywhere else but a corner of the surgical pen, can lead to a wide range of movements by the animal during the operation, which in turn can lead to it becoming recumbent or inflicting inadvertent visceral trauma. Visceral trauma is a result of poor surgical technique or poor patient restraint. Advancing sharp objects, such as trocars or needles intraperitoneally, should be carried out under endoscopic observation. Otherwise, there is a considerable risk of any of the peritoneal viscera becoming traumatized or perforated, with catastrophic consequences for the patient. When inserting trocars or exteriorizing Spieker needles, care should be taken to ensure no damage to the milk vein or its tributaries is caused at the abomasopexy point. Recumbency is usually the result of poor restraint or a higher than required sedative dose. This complication is very rare and, provided there were no sharp objects *in situ* when it occurred, the prognosis is very good.

Uncomplicated cases and those without concurrent disease will recover and return to normal yields within 10 days. The speed of recovery of all other cases depends on the level of complications and nature of concurrent disease.

7 Corrective Laparoscopy – Right Displacement of Abomasum

Abstract

This chapter describes applications of endoscopy in cattle in order to correct the right displacement of abomasum (RDA) and its variations, including the volvulus. RDA is a complicated condition, which very often results in severe difficulties that stem from the stasis of the abdominal content or even the strangulating effects of a volvulus. Prognosis is often grave. This surgical technique allows the surgeon to assess and treat the level of complications with minimal intrusion.

Right displacement of abomasum (RDA), which includes the distension and volvulus of the organ, is a serious condition affecting cattle of all ages, breeds, stages of lactation and management systems, and is very often accompanied by poor prognosis. Decreased appetite, reduced milk production, abdominal distension, particularly on the right-hand side, and abnormal bowel sounds all indicate that a cow may have an RDA. The distinct 'ping' and 'slosh' can be heard at the right paralumbar fossa and right intercostal area during simultaneous auscultation with a stethoscope and percussion, the latter by using the surgeon's fingers.

In cases of abomasal volvulus (AV), the right-hand-side distended abomasum in most cases twists twice, first on its sagittal axis and subsequently on its vertical axis. It is the second twist that is responsible for most of the grave symptoms that accompany this serious condition. These clinical findings range from acute symptoms of hypovolemic shock and toxaemia, to eventual death if prompt intervention has not taken place.

Whenever a case of RDA with or without AV is presented, any possible treatment must be balanced against the prognosis of each case and the financial implications of the chosen treatment. In my opinion, a larger number of RDA are elected to be voluntarily culled on humane grounds, instead of subjected to any treatment, compared to those with left-displaced abomasum (LDA), which are commonly treated.

© Sotirios Karvountzis 2023. *Bovine Endoscopy* (Sotirios Karvountzis)
DOI: 10.1079/9781789246681.0007

Endoscopic correction of RDA and AV presents further diagnostic and prognostic options to the surgeon compared to conventional diagnostic or surgical methods. The procedure for an RDA is always based on the two-step approach, with the first step being the exploratory and preparatory step. The second finds the animal in dorsal recumbency and is the one in which correction takes place.

In cases of AV, two two-step operations 12 hours apart are recommended. The first two-step approach is aimed at identifying the RDA and making an assessment of whether or not there is an AV. When an AV is identified and the decision to proceed with treatment is granted, abomasocentesis, insertion of the abomasal toggle and deflation of the distended abomasum takes place, but without abomasopexy. Abomasopexy will take place during the second two-step operation, after the patient has received a large volume of fluids during the interval between the two two-step operations.

The explanation for the abomasopexy being deferred for the second two-step operation is the large volume of fluid rehydration therapy that the patient receives, which will help bulk up the now reduced abomasum. It will also assist in the passage of the putrid abomasal content that was under stasis during the RDA and the strengthening of the patient. The combination of two two-step operations will improve the chances of recovery.

The restraint of the patient, preparation of the surgical fields and anaesthesia for the two-step RDA corrective technique are the same as with the two-step LDA corrective technique. Likewise surgical and endoscopic instruments used for the RDA correction are the same as those utilized for an LDA two-step correction. Bearing in mind the already elaborated steps in previous chapters on how to carry out the two-step technique, a brief summary of the process follows.

While the animal is standing, commence the first step by accessing the peritoneum via the left paralumbar fossa and set up the optical portal, as described in Chapter 5, Exploratory Laparoscopy. Set up the optical portal, carry out an exploration of the dorsal left abdomen and, if all clear, proceed to the passive or active insufflation of the peritoneal cavity.

When the insufflation is completed, continue with placing the animal in dorsal recumbency. The second step of the two-step operation starts with the establishment of the optical portal. The location of it can be at the animal's left ventral side, a hand's width caudal from the xyphoid process or a hand's width cranial from the navel. The optical portal can also be considered at a hand's width caudal from the xyphoid process on the animal's right-hand ventral side.

With the abdomen already insufflated, consider further insufflation if necessary. Further exploration of the ventral abdomen takes place and, if these findings permit it, identify the distended or volvulated abomasum. In RDA cases, the abomasum lies on the animal's right-hand cranio-ventral abdomen. Furthermore, in order to distinguish between volvulated and non-volvulated cases, the position of the pylorus must be ascertained. In non-volvulated cases, the pylorus lies in its normal position, i.e. pointing caudally. In AV cases, due to the numerous twists that accompany the case, the pylorus points cranially.

In non-volvulated RDA cases, proceed with the abomasocentesis, insertion of the abomasal toggle and deflation of the organ. Once these steps are complete, carry out the abomasopexy while selecting, as a fixation point, a point that lies cranially to the navel area on the animal's right-hand side. The working portal should be set up there.

In AV cases, proceed as described above, but do not carry out an abomasopexy. Once the abomasum is deflated and with the toggles inserted in it, bring the patient into ventral recumbency and proceed with oral rehydration therapy. The exact volume of fluids required depends on the patient's level of dehydration and accompanying toxicity. I have supplied mild AV cases with 40–60 l of appropriately prepared electrolyte solutions over a period of 10–12 hours. In severe cases this may rise to 100 l. The surgeon has to exercise his or her judgement on the matter. The route of administration can be oral or intravenous.

Once the rehydration is complete, the second two-step technique commences, with the aim of carrying out the abomasopexy. Once the second exploration of the ventral abdomen is complete and the findings are satisfactory, carry out the fixation of the fourth stomach as described for the non-AV cases.

Post-operatively, a patient that recovers from an RDA, with or without AV, requires the same supportive therapy as with previously described techniques.

8 Exploratory Thoracoscopy

Abstract

The exploratory thoracoscopy section considers the applications of endoscopy in investigating pulmonary problems. Thoracoscopy is complementary to ultrasonography of the lung and can supplement ultrasound scanning of this region of the body. This is because the surgeon is able to inspect more clearly and further than is possible with other diagnostic methods.

Exploration of the thorax takes place while the patient is standing. The surgeon should consider pneumonic aspiration, which can be carried out with the aid of a two-way fluid pump. A surgical assistant is required to operate the pump, which is connected with a silicon tube to the valve of the magnetic cannula of the optical portal. The assistant operates the pump slowly and it constantly draws air out of the thorax while the procedure is progressing. The purpose of the aspiration is to prevent pneumothorax due to possible air leakage caused by the presence of the optical portal.

The optical portals may lie either on the left- or the right-hand side of the patient. They are located between the 11th and 9th intercostal spaces, halfway down on the thorax. Care needs to be applied in selecting the most appropriate location for the portal, which depends on the size of the patient. A selection of portals for thoracoscopy are shown in Fig. 8.1.

© Sotirios Karvountzis 2023. *Bovine Endoscopy* (Sotirios Karvountzis)
DOI: 10.1079/9781789246681.0008

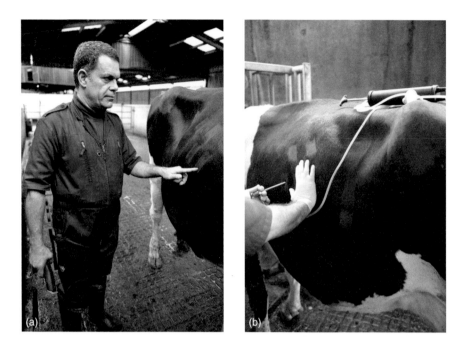

Fig. 8.1. Possible locations for the optical portal in a thoracoscopy. Portals for other endoscopic procedures are also prepared, for illustration purposes. In (b), a 5-mm magnetic trocar is used to set up the portal, aiming to use the 5-mm rigid endoscope. The same image also shows the use of an aspiration pump, which helps minimize the risk of pneumothorax. (Copyright Dr Sotirios Karvountzis.)

The highest risk when thoracoscopy is carried out is the unintended induction of pneumothorax. From the size of the skin incision of the thoracoscopy portal, to the selection of the trocar used, there are very important considerations pertaining to the procedure. The author recommends using the 5-mm magnetic trocar, with the vent of the gas valve set to 'off' while the insertion of the trocar is taking place. This will minimize the risk of unexpected insufflation should the valve be left in the 'on' position while this optical portal is set up.

When inserting the magnetic trocar and cannula through the thoracoscopy optical portal, the surgeon needs to consider the risks of damage to the thoracic organs. It is advisable to avoid entry of the trocar at a right angle to the thorax, but instead a caudo-axial routing should be selected. This aims for the retrothoracic space, while avoiding any damage to the diaphragm. The proposed direction of entry is shown in Fig. 8.1b.

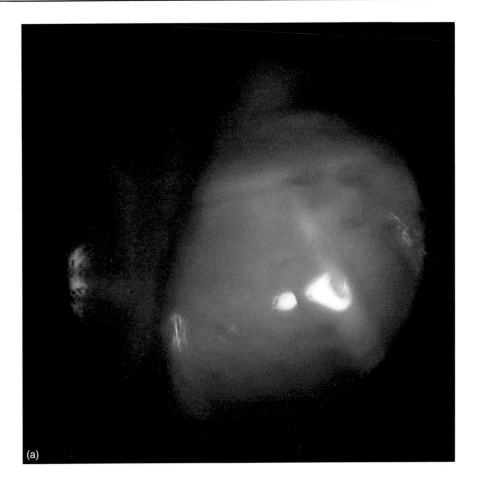

(a)

Finally, an aspiration pump is a useful means to preventing pneumothorax during the procedure. It is a manually operated pump, although electric models based on modified insufflation pumps can also be used. The manual model, as shown in Fig. 8.1b, is operated by an assistant, who shadows the thoracoscopic surgeon and operates it while the procedure is ongoing.

Fig. 8.2 shows an image taken during an exploratory laparoscopy of a dead animal in order to establish the likely cause of illness. The animal was euthanized on humane grounds and the necropsy followed shortly after, before any post-mortem changes affected the carcass. The optical portal was set up on the right-hand side of the chest at the halfway point up the 8th intercostal space. The images depict the medial lobe of the right lung and the heart. No significant findings were found to the heart, but the medial lobe of the right lung was severely congested, a finding consistent with pneumonia. A normal lung should appear pink in colour.

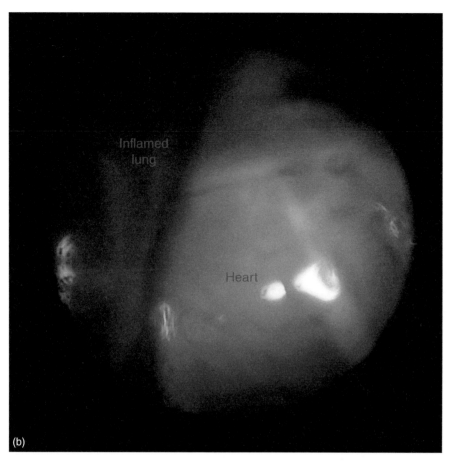

Fig. 8.2. Exploratory laparoscopy during a necropsy, showing a congested medial lobe of the right lung and a healthy heart. (Copyright Dr Sotirios Karvountzis.)

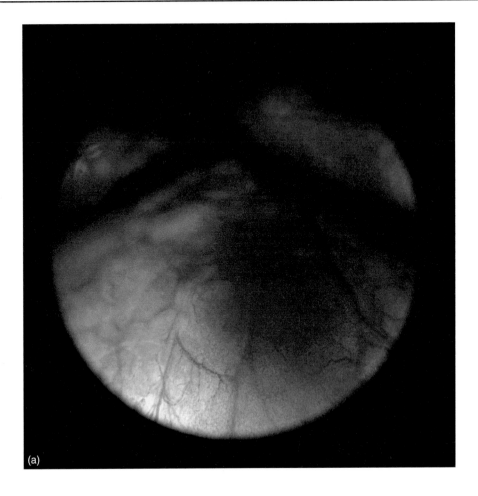

(a)

Fig. 8.3 shows are images taken during an exploratory laparoscopy of a dead animal in order to ascertain the cause of illness. They show a healthy caudal lobe of the left lung. The colour of the lung tissue is pink, indicating it is healthy. The image was taken from the 10th intercostal space, with an optical portal placed on the halfway point between dorsal and ventral.

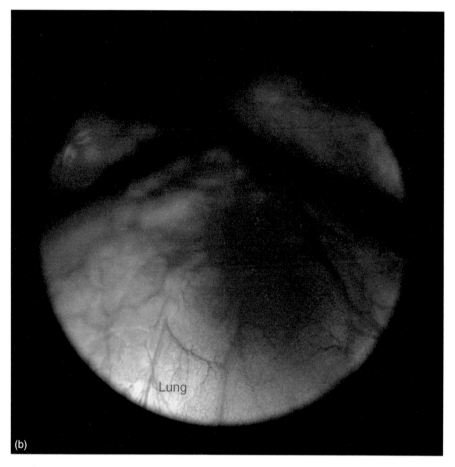

Fig. 8.3. Exploratory laparoscopy during a necropsy, showing a healthy caudal lobe of the left lung. (Copyright Dr Sotirios Karvountzis.)

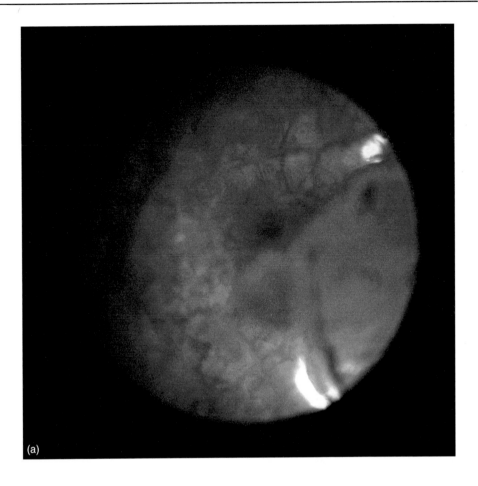

(a)

Fig. 8.4 shows images taken during an exploratory laparoscopy of a dead animal in order to ascertain the cause of illness. They show a healthy caudal lobe of the right lung. The colour of the lung tissue is pink, implying a healthy blood supply to the lung. The image was taken from the 10th intercostal space, with an optical portal placed on the halfway point between dorsal and ventral.

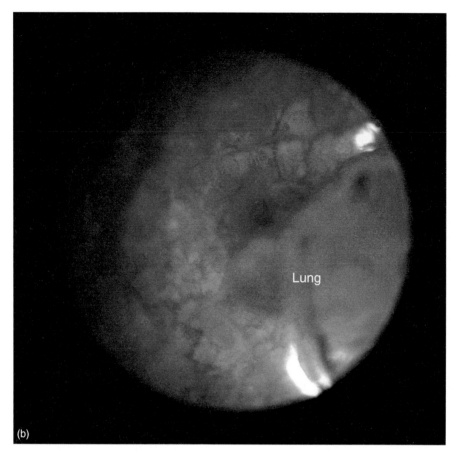

Fig. 8.4. Exploratory laparoscopy during a necropsy, showing a healthy caudal lobe of the right lung. (Copyright Dr Sotirios Karvountzis.)

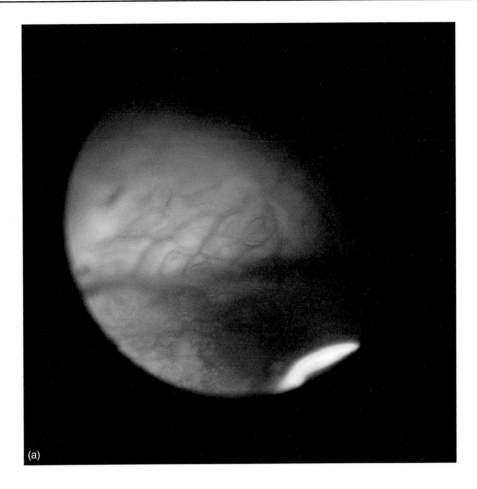

(a)

Fig. 8.5 shows images taken during an exploratory laparoscopy of a dead animal in order to ascertain the cause of illness. They demonstrate a congested medial lobe of the right lung and the adjacent pericardium. The colour of the lung tissue is dark red, implying congestion due to inflammation of the lung. The image was taken from the 8th intercostal space, with an optical portal placed on the halfway point between dorsal and ventral.

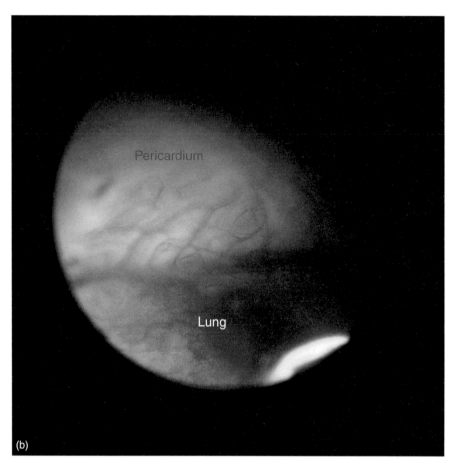

Fig. 8.5. Exploratory laparoscopy during a necropsy, showing a congested medial lobe of the right lung. (Copyright Dr Sotirios Karvountzis.)

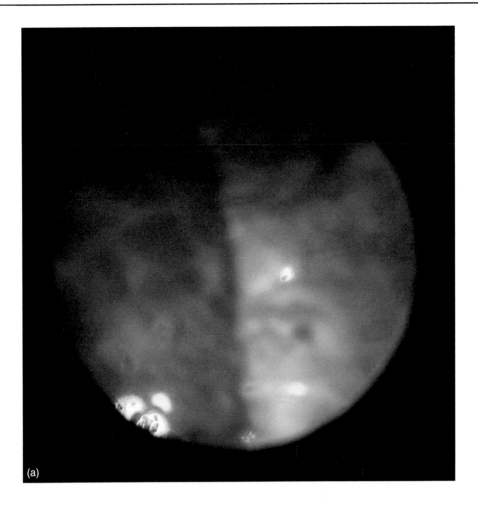

(a)

Fig. 8.6a and b are also images taken during an exploratory laparoscopy of a dead animal in order to ascertain the cause of illness. They demonstrate a congested caudal lobe of the right lung and a healthy medial lobe of the same lung. The colour of the caudal lobe is dark red, indicating inflammation of this lobe. The image was taken from the 8th intercostal space, with an optical portal placed on the halfway point between dorsal and ventral.

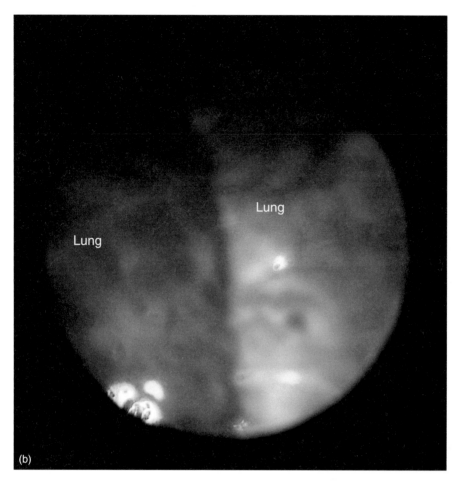

Fig. 8.6. Exploratory laparoscopy during a necropsy, depicting a congested caudal lobe of the right lung and a healthy medial lobe of the same side. (Copyright Dr Sotirios Karvountzis.)

9 Theloscopy

Abstract

This chapter investigates the endoscopic applications of the bovine teat. Bovine theloscopy offers a precision approach to diagnosing and treating teat problems. From teat sphincter constrictions to foreign bodies in the teat canal, most can be approached with this minimally intrusive technique. The outcome is a patient that quickly returns to expected milk yield and reproductive performance, with few complications.

Theloscopy, also known as teat endoscopy, offers a precision approach for diagnosing and treating bovine teat problems. The word theloscopy is derived from the Greek words *thili* (Θηλή) for 'teat', and *skopein* (Σκοπεῖν), for 'inspect'.

With comparatively less risk and inconvenience to the patient compared to conventional methods for diagnosis, treatment and post-treatment monitoring, theloscopy is now an established technique that has been used by veterinarians on hundreds of cows, for procedures ranging from teat sphincter constrictions to foreign bodies in the teat canal. The outcome is a cow that can quickly return to expected milk yield and reproductive performance, with few complications. Post-treatment, cows can match the yield of others in the herd and match their productive longevity.

Theloscopy is often used to diagnose milk flow disorders; for example, it enables the cistern to be visualized during surgery and the response to treatment monitored. Before the advent of theloscopy, precisely diagnosing and treating milk flow disorders with undamaged skin had been challenging. It could have meant inspecting, palpating, probing blindly or milking by hand the affected teat, but often ultrasonography was needed to make a definitive diagnosis.

Teat endoscopy is also an excellent and quick diagnostic procedure for hidden teat injuries. It can provide the exact condition of the mucosa, the intensity or grade and duration of pathological changes. It is minimally invasive

© Sotirios Karvountzis 2023. *Bovine Endoscopy* (Sotirios Karvountzis)
DOI: 10.1079/9781789246681.0009

and allows for the treatment of injuries after diagnosis and the monitoring of treatment. Minimum invasion is the preferred approach for diagnostic procedures involving the teats and udder of large animals.

It is commonly used for the following:

Teat obstruction: The cause of a teat obstruction or blockage, perhaps scar tissue, tumours or a foreign body, can be established with teat endoscopy. The veterinary surgeon will be able to see inside the teat canal, identify the obstruction and determine suitable treatment.

Teat lesions: Cattle can develop papillomas or polyps, teat lesions or growths, which often interfere with milk flow and cause the animal discomfort. Veterinarians utilize teat endoscopy to assess the type of lesion and its magnitude, allowing treatment to be targeted or, if necessary, for the lesion to be removed with surgery.

Teat abnormalities: Investigating congenital or acquired abnormalities, such as clefts or constrictions, is another valuable use for endoscopy because it can allow the severity of the abnormality to be established and, once this has been done, the veterinary surgeon can advise on appropriate management strategies.

Teat surgery: If surgery is required, teat endoscopy can help guide that procedure. For instance, if teat canal surgery is needed, the surgeon can use the endoscope to visualize the surgical site to make the incisions and repairs accurately.

Teat endoscopy is typically carried out in the following order:

Preparation: Having all the necessary equipment to hand is important before getting started on the procedure. The equipment needed will include a teat endoscope, cleaning solutions, sterile lubricating gel and appropriate personal protective equipment.

Restraint: The cow must be secured in a safe and comfortable position, perhaps with a chute or other equipment suitable for keeping her still and calm. All the considerations of the standing technique should be taken into account while safeguarding the theloscopic surgeon from being struck by the patient's rear limbs. This can be achieved by tying a restraining rope on the leg on the same side as the affected teat.

Cleaning and anaesthetizing: Use a mild antiseptic solution to clean the exterior of the teat thoroughly to reduce the risk of introducing infection. Local anaesthetic infiltration is my preferred method. The anaesthetic preparations are the same for theloscopy as they are for laparoscopy. Local infusion is at the base of the teat at between four and six sites, 5 ml each, as shown in Fig. 9.1.

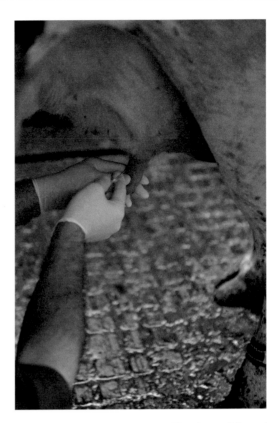

Fig. 9.1. Exploratory theloscopy local anaesthetic infiltration, while using a 5-ml syringe and a 1-inch-long 16-gauge needle to infuse local anaesthetic at the base of the teat. (Copyright Dr Sotirios Karvountzis.)

Insertion: Introduce the endoscope into the teat canal gently, applying steady pressure to carefully manoeuvre it deeper into the teat. Theloscopy can be performed through the lateral teat wall or the teat canal. If the teat wall method is used, it requires a small opening for endoscopic examination. This is sutured after the surgery is complete. When endoscopic examination is performed through the teat canal, that canal and the teat cistern can be inspected through the theloscope in an upward direction, and this is known as axial theloscopy. The view is directed downwards when the approach is through the lateral teat wall, and this is known as lateral theloscopy. With this option, both the teat cistern and the inner opening of the teat canal can be clearly visualized. For the most common disorders in the area of the inner opening of the teat canal, visualization via the lateral wall may be better than via the teat canal; it is a safe and effective method for monitoring the teat after treatment without damaging the teat.

Fig. 9.2. Setting up the optical portal for exploratory theloscopy, using a 5-mm magnetic trocar. (Copyright Dr Sotirios Karvountzis.)

Visualization: The interior of the teat canal will be visualized on a monitor connected to the endoscope as it advances into the teat, when abnormalities such as strictures, lesions or obstructions can be noted.

Examination and sampling: The teat endoscope can be used to take samples such as tissue or milk for further analysis.

Removal: When the examination has been completed, the endoscope should be removed slowly and gently, while being careful to not cause injury or discomfort to the cow.

Post-procedure care: The cow's teat must be kept clean and dry after the procedure and any treatment or medication based on the findings of the examination administered.

The following provides a visual guide to some of the steps involved in teat endoscopy. The optical portal is set up at the side of the teat, halfway down its length, as shown in Fig. 9.2. Once we are ready to commence the theloscopy, the base of the teat is clamped with suitable forceps. The purpose of this step is to stop the insufflated air from escaping into the mammary gland attached to the teat being operated on. The working portal is usually the exiting teat canal. This technique is minimally intrusive, while also permitting full visualization of the teat cistern.

Fig. 9.3. A 3-mm wide and 20-cm long rigid endoscope used in theloscopy procedures, (a) without and (b) with the portable light source. (Copyright Dr Sotirios Karvountzis.)

The theloscope used throughout this publication is a 3-mm wide and 20-cm long rigid endoscope with a screw-on portal for the portable light source. This theloscope is shown in Fig. 9.3.

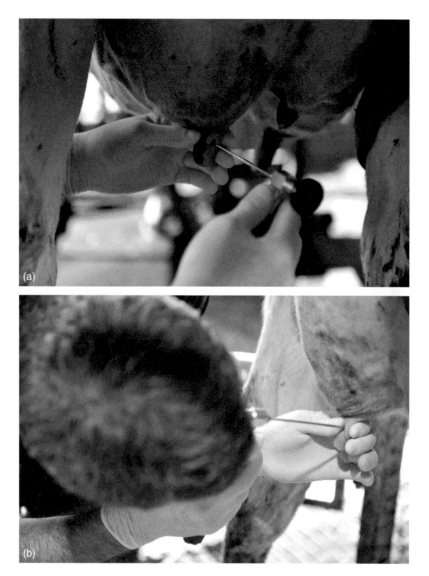

Fig. 9.4. Provided the nearest limb is completely immobilized and no risk to the surgeon is posed, these images show how to utilize the theloscope and examine the offending teat. (Copyright Dr Sotirios Karvountzis.)

When carrying out theloscopic operations, it is paramount that the rear limb that is nearest to the examined teat is stabilized and restrained. The surgeon can lift the cow's leg, or can tie the leg to the ground, so that the cow cannot lift it. The importance of protecting the surgeon and safeguarding the animal that is undergoing theloscopy cannot be stressed enough.

Depending on which teat is to be examined, the theloscopic portal is set up on the cranial or caudal aspect of the teat wall. Ideally, the theloscopic portal is located a finger's width ventrally to the base of the teat, so that it provides adequate clearance from the teat base or the teat sphincter. The location of the theloscopic portal is shown in Fig. 9.4.

It is preferrable that insufflation takes place into the teat cistern, so that the internal walls of the teat will be pushed apart and allow the theloscopic surgeon the extra room to guide his or her instruments intra-thelially. Very often, depending on the conditions in the field, it is not possible to carry the insufflation pump with us in order to introduce air into the cavity. In such cases, although the working space is drastically reduced, it is still possible to carry out theloscopic operations successfully.

References

Babkine, M. and Desrochers, A. (2005) Laparoscopic surgery in adult cattle. *Veterinary Clinics Food Animal Practice* 21, 251–279.

Babkine, M., Desrochers, A., Bouré, L. and Hélie, P. (2006) Ventral laparoscopic abomasopexy on adult cows. *Canadian Veterinary Journal* 47, 343–348.

Baird, A.N. (2012) Displaced abomasum in cattle: evaluation beyond the ping. *Veterinary Record* 170, 411–412.

Bouré, L. (2005) General principles of laparoscopy. *Veterinary Clinics Food Animal Practice* 21, 227–249.

Christiansen, K. (2000) Laparoskopisch kontrollierte Operation des nach links verlagerten Labmagens (Janowitz Operation) ohne Ablegen des Patienten. *Für Studium und Praxis* 2004, 118–121.

Constable, P.D., Miller, G.Y., Hoffsis, G.F., Hull, B.L. and Rings, D.M. (1992) Risk factors for abomasal volvulus and left abomasal displacement in cattle. *American Journal of Veterinary Research* 53, 1184–1192.

Dawson, L.J., Aalseth, E.P., Rice, L.E. and Adams, G.D. (1992) Influence of fiber form in a complete mixed ration on incidence of left displaced abomasum in postpartum dairy cows. *Journal of the American Veterinary Medical Association* 200, 1989–1992.

Detilleux, J.C., Gröhn, Y.T., Eicker, S.W. and Quaas, R.L. (1997) Effects of left displaced abomasum on test day milk yields of Holstein cows. *Journal of Dairy Science* 80, 121–126.

Devaux, D.J., Lempen, M., Schelling, E., Koch, V.M. and Meylan, M. (2015) Assessment of the excretion time of electronic capsules placed in the intestinal lumen of cows with cecal dilatation dislocation, healthy control cows, and cows with left displacement of the abomasum. *American Journal of Veterinary Research* 76, 60–69.

Doll, K., Sickinger, M. and Seeger, T. (2008) New aspects in the pathogenesis of abomasal displacement. *Veterinary Journal* 181, 90–96.

Fubini, S.L., Ducharme, N.G., Erb, H.N. and Sheils, R.L. (1992) A comparison in 101 dairy cows of right paralumbar fossa omentopexy and right paramedian abomasopexy for treatment of left displacement of the abomasum. *Canadian Veterinary Journal* 33, 318–324.

Grosche, A., Fürll, M. and Wittek, T. (2012) Peritoneal fluid analysis in dairy cows with left displaced abomasum and abomasal volvulus. *Veterinary Record* 170, 413–418.

Grymer, J. and Sterner, K.E. (1982) Percutaneous fixation of left displaced abomasum, using a bar suture. *Journal of the American Veterinary Medical Association* 180, 1458–1461.

Hamann, H., Wolf, V., Scholz, H. and Distl, O. (2004) Relationships between lactational incidence of displaced abomasum and milk production traits in German Holstein cows. *Journal of Veterinary Medicine* 51, 203–208.

Itoh, N., Egawa, M., Kitazawa, T., Ueda, M. and Koiwa, M. (2006) A new method for detecting the abomasal position and characteristics of movement at the onset of the left displacement of the abomasum in cows. *Journal of Veterinary Medicine* 53, 375–378.

Janowitz, H. (1998) Laparoskopische Reposition und Fixation des nach links verlagerten Labmagens beim Rind. *Tierärtzliche Praxis* 26, 308–313.

Jorritsma, R., Westerlaan, B., Bierma, M.P.R. and Frankena, K. (2008) Milk yield and survival of Holstein-Friesian dairy cattle after laparoscopic correction of left-displaced abomasum. *Veterinary Record* 162, 743–746.

Karvountzis, S. (2018) Left displaced abomasum: pilot survey of corrective techniques. *Veterinary Record* 178, 509–510.

Karvountzis, S. (2020) Splenoptosis in a dairy cow and endoscopic correction of left displacement of abomasum. *Veterinary Record Case Reports* 8, https://doi.org/10.1136/vetreccr-2020-001135.

Lee, I., Yamagishi, N., Oboshi, K. and Yamada, H. (2002) Left paramedian abomasopexy in cattle. *Journal of Veterinary Science* 3, 59–60.

Martin, S.W., Kirby, K.L. and Curtis, R.A. (1978a) A study of the role of genetic factors in the etiology of left abomasal displacement. *Canadian Journal of Comparative Medicine* 42, 511–518.

Martin, S.W., Kirby, K.L. and Curtis, R.A. (1978b) Left abomasal displacement in dairy cows: its relationship to production. *Canadian Veterinary Journal* 19, 250–253.

Martin, W. (1972) Left abomasal displacement: an epidemiological study. *Canadian Veterinary Journal* 13, 61–68.

McArt, J.A.A., Nydam, D.V. and Oetzel, G.R. (2012) Epidemiology of subclinical ketosis in early lactation dairy cattle. *Journal of Dairy Science* 95, 5056–5066.

Mees, K. (2011) Effects of a cow drink on the convalescence of dairy cows after surgical correction of left abomasal displacement using the 'roll & toggle' approach according to Grymer and Sterner. Doctoral thesis, Department of Veterinary Medicine, University of Berlin Germany.

Mulon, P.Y., Babkine, M. and Desrochers, A. (2006) Ventral laparoscopic abomasopexy in 18 cattle with displaced abomasum. *Veterinary Surgery* 35, 347–355.

Newman, K.D., Anderson, D.E. and Silveira, F. (2005) One-step laparoscopic abomasopexy for correction of left-sided displacement of the abomasum in dairy cows. *Journal of the American Veterinary Medical Association* 227, 1142–1147.

Newman, K.D., Harvey, D. and Roy, J.P. (2008) Minimally invasive field abomasopexy: techniques for correction and fixation of left displacement of the abomasum in dairy cows. *Veterinary Clinics Food Animal Practice* 24, 359–382.

Ospina, P.A., Nydam, D.V., Stokol, T. and Overton, T.R. (2009) Associations of elevated non-esterified fatty acids and β-hydroxybutyrate concentrations with early lactation reproductive performance and milk production in transition dairy cattle in the northeastern United States. *Journal of Dairy Science* 93, 1596–1603.

Rager, K.D., George, L.W., House, J.K. and DePeters, E.J. (2004) Evaluation of rumen transfaunation after surgical correction of left-sided displacement of the abomasum in cows. *Journal of the American Veterinary Medical Association* 225, 915–920.

Raizman, E.A. and Santos, J.E.P. (2002) The effect of left displacement of abomasum corrected by toggle-pin suture on lactation, reproduction and health of Holstein dairy cows. *Journal of Dairy Science* 85, 1157–1164.

Rohn, M., Tenhagen, B.A. and Hofmann, W. (2004) Survival of dairy cows after surgery to correct abomasal displacement: clinical and laboratory parameters and overall survival. *Journal of Veterinary Medicine* 51, 294–299.

Roussel, A.J., Cohen, N.D. and Hooper, R.N. (2000) Abomasal displacement and volvulus in beef cattle: 19 cases (1988–1998). *Journal of the American Veterinary Medical Association* 216, 730–733.

Roy, J.P., Harvey, D., Bélanger, A.M. and Buczinski, S. (2008) Comparison of 2-step laparoscopy-guided abomasopexy versus omentopexy via right flank laparotomy for the treatment of dairy cows with left displacement of the abomasum in on-farm settings. *Journal of the American Veterinary Medical Association* 232, 1700–1706.

Seeger, T., Kümper, H., Failing, K. and Doll, K. (2006) Comparison of laparoscopic guided abomasopexy versus omentopexy via right flank laparotomy for the treatment of left abomasal displacement in dairy cows. *American Journal of Veterinary Research* 67, 472–478.

Sexton, M.F., Buckley, W. and Ryan, E. (2007) A study of 54 cases of left displacement of the abomasum: February to July 2005. *Irish Veterinary Journal* 60, 605–609.

Shaver, R.D. (1997) Nutritional risk factors in the etiology of left displaced abomasum in dairy cows: a review. *Journal of Dairy Science* 80, 2449–2453.

Sobiech, P., Radwińska, J., Krystkiewicz, W., Snarska, A. and Stopyra, A. (2008) Changes in the coagulation profile of cattle with left abomasal displacement. *Polish Journal of Veterinary Sciences* 11, 301–306.

Sterner, K. and Grymer, J. (2002) 20 years' experience with the Grymer/Sterner® toggle suture technique for LDA repair: improvements in materials and methods. Paper presented at XXII World Buiatrics Congress, Germany.

Steven, C.L., Van Winden, S.C. and Kuiper, R. (2003) Left displacement of the abomasum in dairy cattle: recent developments in epidemiological and etiological aspects. *Veterinary Research* 34, 47–56.

Van Winden, S.C., Brattinga, C.R., Müller, K.E., Schonewille, J.T., Noordhuizen, J.P. and Beynen, A.C. (2004) Changes in the feed intake, pH and osmolality of rumen fluid, and the position of the abomasum of eight dairy cows during a diet-induced left displacement of the abomasum. *Veterinary Record* 154, 501–504.

Wittek, T., Fürll, M. and Grosche, A. (2012) Peritoneal inflammatory response to surgical correction of left displaced abomasum using different techniques. *Veterinary Record* 171, 594–599.

Wittek, T., Schreiber, K., Fürll, M. and Constable, P.D. (2005) Use of the D-Xylose absorption test to measure abomasal emptying rate in healthy lactating Holstein-Friesian cows and in cows with left displaced abomasum or abomasal volvulus. *Journal of Veterinary Internal Medicine* 19, 905–913.

Wittek, T., Tischer, K., Giesler, T., Fürll, M. and Constable, P.D. (2008) Effect of preoperative administration of erythromycin or flunixin meglumine on postoperative abomasal emptying rate in dairy cows undergoing surgical correction of left displacement of the abomasum. *Journal of the American Veterinary Medical Association* 232, 418–423.

Zadnik, T. and Lombar, R. (2011) Our experience with left-sided abomasal displacement correction via the roll-and-toggle-pin suture procedure according to Grymer/Sterner model. *Veterinary Science* 2011, 1–3.

Zerbin, I., Lehner, S. and Distl, O. (2015) Genetics of bovine abomasal displacement. *Veterinary Journal* 204, 17–22.

Index

Note: The page numbers in *italics* refer to figures.